HANDBOOK OF SPUTTER DEPOSITION TECHNOLOGY

HANDBOOK OF SPUTTER DEPOSITION TECHNOLOGY

Principles, Technology and Applications

by

Kiyotaka Wasa

Matsushita Electric Industrial Co., Ltd.
Osaka, Japan

Shigeru Hayakawa

Ion Engineering Center Corporation
Osaka, Japan

np | **NOYES PUBLICATIONS**
Park Ridge, New Jersey, U.S.A.

Library of Congress Catalog Card Number: 90-27820
ISBN: 0-8155-1280-5
Printed in the United States

Published in the United States of America by
Noyes Publications
Mill Road, Park Ridge, New Jersey 07656

10 9 8 7 6 5 4 3 2 1

Library of Congress Cataloging-in-Publication Data

Wasa, Kiyotaka.
        Handbook of sputter deposition technology : principles,
technology, and applications / by Kiyotaka Wasa and Shigeru
Hayakawa.
            p.   cm.
        Includes bibliographical references and index.
        ISBN 0-8155-1280-5 :
        1. Cathode sputtering (Plating process) 2. Thin films.
I. Hayakawa, Shigeru, 1925-   . II. Title.
TS695.W37   1991
621.3815'2--dc20   90-27820
                                    CIP

MATERIALS SCIENCE AND PROCESS TECHNOLOGY SERIES

Editors

Rointan F. Bunshah, University of California, Los Angeles *(Series Editor)*

Gary E. McGuire, Microelectronics Center of North Carolina *(Series Editor)*

Stephen M. Rossnagel, IBM Thomas J. Watson Research Center *(Consulting Editor)*

Electronic Materials and Process Technology

DEPOSITION TECHNOLOGIES FOR FILMS AND COATINGS: by Rointan F. Bunshah et al

CHEMICAL VAPOR DEPOSITION FOR MICROELECTRONICS: by Arthur Sherman

SEMICONDUCTOR MATERIALS AND PROCESS TECHNOLOGY HANDBOOK: edited by Gary E. McGuire

HYBRID MICROCIRCUIT TECHNOLOGY HANDBOOK: by James J. Licari and Leonard R. Enlow

HANDBOOK OF THIN FILM DEPOSITION PROCESSES AND TECHNIQUES: edited by Klaus K. Schuegraf

IONIZED-CLUSTER BEAM DEPOSITION AND EPITAXY: by Toshinori Takagi

DIFFUSION PHENOMENA IN THIN FILMS AND MICROELECTRONIC MATERIALS: edited by Devendra Gupta and Paul S. Ho

HANDBOOK OF CONTAMINATION CONTROL IN MICROELECTRONICS: edited by Donald L. Tolliver

HANDBOOK OF ION BEAM PROCESSING TECHNOLOGY: edited by Jerome J. Cuomo, Stephen M. Rossnagel, and Harold R. Kaufman

CHARACTERIZATION OF SEMICONDUCTOR MATERIALS—Volume 1: edited by Gary E. McGuire

HANDBOOK OF PLASMA PROCESSING TECHNOLOGY: edited by Stephen M. Rossnagel, Jerome J. Cuomo, and William D. Westwood

HANDBOOK OF SEMICONDUCTOR SILICON TECHNOLOGY: edited by William C. O'Mara, Robert B. Herring, and Lee P. Hunt

HANDBOOK OF POLYMER COATINGS FOR ELECTRONICS: by James J. Licari and Laura A. Hughes

HANDBOOK OF SPUTTER DEPOSITION TECHNOLOGY: by Kiyotaka Wasa and Shigeru Hayakawa

HANDBOOK OF VLSI MICROLITHOGRAPHY: edited by William B. Glendinning and John N. Helbert

CHEMISTRY OF SUPERCONDUCTOR MATERIALS: edited by Terrell A. Vanderah

CHEMICAL VAPOR DEPOSITION OF TUNGSTEN AND TUNGSTEN SILICIDES: by John E.J. Schmitz

(continued)

Ceramic and Other Materials—Processing and Technology

SOL-GEL TECHNOLOGY FOR THIN FILMS, FIBERS, PREFORMS, ELECTRONICS AND SPECIALTY SHAPES: edited by Lisa C. Klein

FIBER REINFORCED CERAMIC COMPOSITES: by K.S. Mazdiyasni

ADVANCED CERAMIC PROCESSING AND TECHNOLOGY—Volume 1: edited by Jon G.P. Binner

FRICTION AND WEAR TRANSITIONS OF MATERIALS: by Peter J. Blau

SHOCK WAVES FOR INDUSTRIAL APPLICATIONS: edited by Lawrence E. Murr

SPECIAL MELTING AND PROCESSING TECHNOLOGIES: edited by G.K. Bhat

CORROSION OF GLASS, CERAMICS AND CERAMIC SUPERCONDUCTORS: edited by David E. Clark and Bruce K. Zoitos

Related Titles

ADHESIVES TECHNOLOGY HANDBOOK: by Arthur H. Landrock

HANDBOOK OF THERMOSET PLASTICS: edited by Sidney H. Goodman

SURFACE PREPARATION TECHNIQUES FOR ADHESIVE BONDING: by Raymond F. Wegman

FORMULATING PLASTICS AND ELASTOMERS BY COMPUTER: by Ralph D. Hermansen

Preface

Cathodic sputtering is currently being widely used in the microelectronics industry for the production of silicon integrated circuits. Historically, cathodic sputtering was first observed in the 1800's, but has developed rapidly over the past 20 years for applications for microelectronics and metallurgical coatings. Recently interest has again increased in cathodic sputtering since the novel materials, i.e., high-temperature superconductors, can be synthesized with sputtering under nonthermal equilibrium conditions.

Several books have described sputtering phenomena and applications for a deposition of thin films. For example, two early books, Thin Film Phenomena, by K.L. Chopra (1969) and the Handbook of Thin Film Technology edited by L. Maissel and G. Glang (1970), provided excellent reviews of this field about 20 years ago. Two text books, Sputtering by Particle Bombardment I & II, edited by R. Behrish (1981, 1983), completely review the recent scientific studies on sputtering. More recently, the Handbook of Ion Beam Processing Technology, edited by J.J. Cuomo, S.M. Rossnagel and H.R. Kaufman (1989), and the Handbook of Plasma Processing Technology, edited by S.M. Rossnagel, J.J. Cuomo and W.D. Westwood (1990) review recent developments in plasma and sputtering technology.

However, a concise and organized textbook on sputtering and sputter deposition technology is still desired as a valuable resource for graduate students and workers in the field.

The authors have studied cathodic sputtering and the sputter deposition of thin films for over 25 years. This book is effectively a comprehensive compilation of the author's works on sputtering technology.

The basic processes relating to thin film materials, growth and deposition techniques are covered in Chapters 1 and 2. The basic concepts of physical sputtering are described in Chapter 3, and the experimental systems used for sputtering applications are described in Chapter 4. A wide range of applications of thin films and deposition technology are described in Chapter

5. The extensive review of physics of the thin film growth in Chapter 2 was contributed by K.L. Chopra (Indian Institute of Technology) to whom I am very grateful. Most of the basic data on the sputtering in Chapter 3 were provided by Professor G.K. Wehner (University of Minnesota). Since the preparation and characterization of sputtered films are both vital parts of sputtering research, Chapter 5 is devoted to these topics, including the author's original works on the sputtering deposition as well as his recent studies on the thin film processing of high-temperature superconductors. The microfabrication of electronic devices and IC's by sputter deposition and related technology is discussed in Chapter 6. In Chapter 7 future directions for sputtering technology are listed. This textbook is intended for use as a reference and research book for graduate students, scientists, and engineers.

I am grateful to Professor G.K. Wehner, who has in his original research on sputtering provided the framework for studies in the field. I am also grateful to Professor R.F. Bunshah (University of California, Los Angeles) and Dr. S.M. Rossnagel (IBM Thomas J. Watson Research Center) for their valuable discussions on this manuscript. I am also grateful to G. Narita (Vice President Executive Editor, Noyes Publications) for his continuous aid and support for the publication of this book. Thanks are due to H. Shano, Y. Wasa and K. Hirochi for typing the manuscripts and preparing most of the illustrations. I acknowledge the assistance of my colleagues in Materials Science Laboratory of Central Research Laboratories, Matsushita Electric Ind. Co. Ltd. I thank A. Tanii, President of Matsushita Electric Ind. Co. Ltd. for his continuous encouragement and support. Finally, this book could not be published without the constant help and understanding of my wife, Setsuko Wasa.

Osaka, Japan   Kiyotaka Wasa
October 1991

NOTICE

Contents

1. **THIN FILM MATERIALS AND DEVICES** 1
   - **1.1** **Thin Film Materials** 1
   - **1.2** **Thin Film Devices** 6
   - **1.3** **References** 8

2. **THIN FILM PROCESSES** 10
   - **2.1** **Thin Film Growth Process** 10
     - 2.1.1 Structural Consequences of the Growth Process 13
       - 2.1.1.1 Microstructure 13
       - 2.1.1.2 Surface Roughness 15
       - 2.1.1.3 Density 17
       - 2.1.1.4 Adhesion 18
       - 2.1.1.5 Metastable Structure 18
     - 2.1.2 Solubility Relaxation 19
   - **2.2** **Thin Film Deposition Process** 19
     - 2.2.1 Classification of Deposition Process 19
     - 2.2.2 Deposition Conditions 29
   - **2.3** **Characterization** 37
   - **2.4** **References** 47

3. **SPUTTERING PHENOMENA** 49
   - **3.1** **Sputter Yield** 49
     - 3.1.1 Ion Energy 50
     - 3.1.2 Incident Ions, Target Materials 54
     - 3.1.3 Angle of Incidence Effects 57
     - 3.1.4 Crystal Structure of Target 59
     - 3.1.5 Sputter Yields of Alloys 61

ix

3.2   Sputtered Atoms 64
      3.2.1   Features of Sputtered Atoms 64
      3.2.2   Velocity and Mean Free Path 65
              3.2.2.1   Velocity of Sputtered Atoms 65
              3.2.2.2   Mean Free Path 70
3.3   Mechanisms of Sputtering 71
      3.3.1   Sputtering Collisions 71
      3.3.2   Sputtering 73
3.4   References 78

4. SPUTTERING SYSTEMS 81
4.1   Discharge in a Gas 81
      4.1.1   Cold-Cathode Discharge 81
      4.1.2   Discharge in a Magnetic Field 88
              4.1.2.1   Spark Voltage in a Magnetic Field 88
              4.1.2.2   Glow Discharge in a Magnetic Field 91
              4.1.2.3   Glow Discharge Modes in the Transverse
                        Magnetic Field 93
              4.1.2.4   Plasma in a Glow Discharge 95
4.2   Sputtering Systems 97
      4.2.1   DC Diode Sputtering 97
      4.2.2   Rf Diode Sputtering 98
      4.2.3   Magnetron Sputtering 100
      4.2.4   Ion Beam Sputtering 105
      4.2.5   ECR Plasmas 106
4.3   Practical Aspects of Sputtering Systems 107
      4.3.1   Targets for Sputtering 107
              4.3.1.1   Compound Targets 110
              4.3.1.2   Powder Targets 111
              4.3.1.3   Auxiliary Cathode 112
      4.3.2   Sputtering Gas 112
      4.3.3   Thickness Distribution 115
      4.3.4   Substrate Temperature 116
      4.3.5   Monitoring 117
              4.3.5.1   Gas Composition 117
              4.3.5.2   Sputtering Discharge 118
              4.3.5.3   Plasma Parameters 119
              4.3.5.4   Thickness Monitor 121
4.4   References 122

5. DEPOSITION OF COMPOUND THIN FILMS 124
5.1   Oxides 125
      5.1.1   ZnO Thin Films 125
              5.1.1.1   Deposition of ZnO 135

		5.1.1.2	Electrical Properties and Applications	147
	5.1.2	Sillenite Thin Films	156
		5.1.2.1	Amorphous, Polycrystal Films	157
		5.1.2.2	Single Crystal Films	160
	5.1.3	Perovskite Dielectric Thin Films	162
		5.1.3.1	$PbTiO_3$ Thin Films	162
		5.1.3.2	PLZT Thin Films	175
	5.1.4	Perovskite Superconducting Thin Films	193
		5.1.4.1	Studies of Thin Film Processes	197
		5.1.4.2	Basic Thin Film Processes	198
		5.1.4.3	Synthesis Temperature	203
		5.1.4.4	Low Temperature Processes/In Situ Deposition	204
		5.1.4.5	Deposition: Rare Earth High T_c Superconductors	205
		5.1.4.6	Deposition: Rare Earth Free High T_c Superconductors	212
		5.1.4.7	Structure and Structural Control	215
		5.1.4.8	Phase Control by Layer-by-Layer Deposition	220
		5.1.4.9	Diamagnetization Properties	221
		5.1.4.10	Passivation of Sputtered High T_c Thin Films	223
	5.1.5	Transparent Conducting Films	226
5.2	**Nitrides**	227
	5.2.1	TiN Thin Films	228
	5.2.2	Compound Nitride Thin Films	228
	5.2.3	SiN Thin Films	230
5.3	**Carbides and Silicides**	231
	5.3.1	SiC Thin Films	231
	5.3.2	Tungsten Carbide (WC) Thin Films	238
	5.3.3	Mo-Si Thin Films	241
5.4	**Diamond**	242
5.5	**Selenides**	245
5.6	**Amorphous Thin Films**	248
	5.6.1	Amorphous ABO_3	250
	5.6.2	Amorphous SiC	253
5.7	**Super-Lattice Structures**	254
5.8	**Organic Thin Films**	256
5.9	**Magnetron Sputtering Under a Strong Magnetic Field**	257
	5.9.1	Abnormal Crystal Growth	258
	5.9.2	Low Temperature Doping of Foreign Atoms into Semiconducting Films	259
5.10	**References**	264

6. MICROFABRICATION BY SPUTTERING 275
   6.1   Ion Beam Sputter Etching 275
   6.2   Diode Sputter Etching 285
   6.3   Plasma Etching 287
   6.4   References 289

7. FUTURE DIRECTIONS 291
   7.1   Conclusions 291
   7.2   References 293

APPENDIX 294
   Electric Units, Their Symbols and Conversion Factors 294
   Fundamental Physical Constants 295

INDEX 296

1

THIN FILM MATERIALS AND DEVICES

Thin solid films are fabricated by the deposition of individual atoms on a substrate. Their thicknesses are typically less than several microns. Historically Bunsen and Grove first obtained thin metal films in a vacuum system in 1852.

Thin films are now widely used for making electronic devices, optical coatings and decorative parts. Thin films are also necessary for the development of novel optical devices, as well as such areas as hard coatings and wear resistant films. By variations in the deposition process, as well as modifications of the film properties during deposition, a range of unusual properties can be obtained which are not possible with bulk materials.

Thin film materials and deposition processes have been reviewed in several publications (1). Among the earlier publications, the "Handbook of Thin Film Technology", edited by Maissel and Glang, is notable although more than 20 years has passed since the book was published and many new and exciting developments have occurred in the intervening years.

1.1 THIN FILM MATERIALS

Thin film materials will exhibit the following special features:

1. Unique material properties resulting from the atomic growth process.
2. Size effects, including quantum size effects characterized by the thickness, crystalline orientation, and multilayer aspects.

Bulk materials are often sintered from powders of source materials. The particle size of these powders is of the order 1μ in diameter. Thin films are synthesized from atoms or small groups of atoms. These ultrafine particles are generally effectively quenched on substrates during film growth, and this non-equilibrium aspect can lead to the formation of exotic materials. A variety of abnormal crystal phases have been reported in thin films. A typical example is the tetragonal Ta reported by Read (2). An amorphous phase can

1

also be observed in thin films which is not characteristic of the bulk material.   Other structures found in the growth of thin films are a island structure of ultra-thin layers or a fiber structure.

Recent progress in thin film growth technology enables one to make novel thin film materials including diamonds and high temperature oxide superconductors. Bulk diamonds are conventionally synthesized at high pressure ($\approx$50,000 psi) and high temperature (2000°C).   The   deposition   of   diamonds   from   energetic   carbon   ions ($\approx$10 $-$ 100eV) enables the growth of the diamond crystallites and/or diamond films at room temperature (3). Thin films of high temperature superconductors are indispensable not only for making thin film superconducting devices but also for studying fundamental aspects of these new superconductors (1). Figure 1.1 shows some photographs of novel thin film materials.

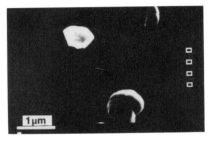

(a) Diamond crystals prepared at room temperature.

(b) Electro-optic thin films of compound (Pb,La)(Zr,Ti)O_3.

(c) High T_c superconducting films on a Si wafer

Figure 1.1: Novel thin materials prepared by cathodic sputtering.

One must also consider that due to abnormal structure accompanied by size effects, thin films may show different features in terms of mechanical strength, carrier transportation, superconducting transitions, magnetic properties, and optical properties. For instance, thin films may be characterized by a strong internal stress of $10^9 - 10^{10}$dynes/cm and a number of lattice defects. The density of the lattice defects can be more than 10^{11} dislocations/cm^2. These lattice defects have the effect of increasing the elastic strength. The strengths obtained in thin films can be up to 200 times as large as those found in corresponding bulk material. Thermal stress arising from the thermal expansion of thin films has been shown to increase the critical temperature of superconducting films (5).

	Increase of resistivity, ρ in metal, $\rho_F/\rho_B \simeq (4/3)(\gamma \ln(1/\gamma))^{-1}$.
SIZE EFFECTS	Reduced TCR, α, in metal, $\alpha_F/\alpha_B \simeq (\ln(1/\gamma))^{-1}$.
$\gamma = t/l << 1$,	Reduced mobility, μ, in metal $\mu_F/\mu_B \simeq (\ln(1/\gamma))^{-1}$.
t: film thickness	Anomalous skin effect at high frequencies in metal.
	Reduced thermal conductivity, K, in metal, $K_F/K_B \simeq (3/4)(\gamma \ln(1/\gamma))$.
l: mean free path of electrons	Enhanced thermoelectric power, S, in metal, $S_F/S_B \simeq 1 + (2/3)(\ln \gamma - 1.42/\ln \gamma - 0.42)$.
	Reduced mobility in semiconductor, $\mu_F/\mu_B \simeq (1 + 1/\gamma)^{-1}$.
	Quantum size effects in semiconductors and semimetal, at t < l, DeBroglie wavelength: thickness-dependant oscillatory variation of resistivity, Hall coefficient, Hall mobility and magnetoresistance. Galvanomagnetic surface effects on Hall effect and magnetoresistance due to surface scattering.

Table 1.1: Interesting phenomena expected in thin film materials. Electron transport phenomena (F = film, B = bulk).

FIELD EFFECTS	Conductance change in semiconductor surface by means of electric field, Insulated-gate thin film transistor (TFT).
SPACE CHARGE LIMITED CURRENT (SCLC)	SCLC through insulator, J: $J = 10^{-13}\mu dEV^2/t^3\ (A/cm^2)$ (one-carrier trap-free SCLS) μd, drift mobility of charge carriers, E, dielectric constant, V, applied voltage
TUNNELING EFFECTS	Tunnel current through thin insulating films, voltage-controlled negative resistance in tunnel diode. Tunnel emission from metal, hot electron triode of metal-base transistor. Electroluminescence, photoemission of electrons. Tunnel spectroscopy. -- Tunnel current between island structure in ultra thin films.
MAGNETICS	Increase in magnetic anisotropy. The anisotropies originate in a shape anisotropy, magnetocrystalline anisotropy, strain-magnetostriction anisotropy, uniaxial shape-anisotropy. Increase in magnetization and permeability in amorphous structure, and/or layered structure.

Table 1.1: (continued) Interesting phenomena expected in thin film materials.

SUPERCON-
DUCTIVITY
	Superconductivity-enhancement:
	  increase of critical temperature, T_c,
	  in metal with decreasing thickness, t,
	  $\Delta T_c \simeq A/t - B/t^2$,
	  and/or crystallite size.

Stress effects:
  tensile stress increases T_c.,
  compressive stress decreases T_c in metal.

Proximity effects in superimposed films:
  decrease of T_c in metal caused by
  contact of normal metal.
Reduced transition temperature, T_t,
  $(T_t/T_c)^2 = 1 - 1/(0.2 + 0.8ts)$,
  ts, ratio of thickness of superconducting films
  and a critical thickness below which no
  superconductivity is observed for a
  constant thickness of normal metal films.

Increase of critical magnetic field, H_c,
  at parallel field,
  $H_{CF}/H_{CB} \simeq \sqrt{24}\ \lambda/t$,
  λ, penetration depth, due to G-L theory.
  at transverse field,
  $H_{CF}/H_{CB} = \sqrt{2}\ K$,
  K, Ginzburg-Landau parameter.

Reduced critical current, J_C,
  $J_{CF}/J_{CB} \simeq \tanh(t/2\lambda)$.
Supercurrent tunneling through thin barrier,
  Josephson junction, and Tunnel spectroscopy.

Table 1.1: (continued) Interesting phenomena expected in thin film materials.

1.2 THIN FILM DEVICES

Since the latter part of the 1950's thin films have been extensively studied in relation to their applications for making electronic devices. In the early 1960's Weimer proposed thin film transistors (TFT) composed of CdS semiconducting films. He succeeded in making a 256-stage thin-film transistor decoder, driven by two 16-stage shift resistors, for television scanning, and associated photoconductors, capacitors, and resistors (7). Although these thin film devices were considered as the best development of both the science and technology of thin films for an integrated microelectronic circuit, the poor stability observed in TFT's was an impediment to practical use. Thus, in the 1960's thin film devices for practical use were limited to passive devices such as thin film resistors and capacitors. However, several novel thin film devices were proposed, including man-made superlattices (8), thin film surface acoustic wave (SAW) devices (9), and thin film integrated optics (10).

In the 1970's a wide variety of thin film devices were developed. Of these, one of the most interesting areas is a thin film amorphous silicon (a-Si) technology proposed by Spear (11). This technology achieved low temperature doping of impurities into a-Si devices and suggested the possibility of making a-Si active devices such as a-Si TFT and a-Si solar cells (12-13). In the 1980's rapid progress was made in a-Si technology. Amorphous Si solar cells have been produced for an electronic calculator although the energy conversion efficiency is 5 to 7% and is lower than that of crystalline Si solar cells. This efficiency, however, has recently been improved (14).

In the middle of the 1980's high quality a-Si technology has led to the production of a liquid crystal television with a-Si TFT. Other interesting thin film devices recently produced are ZnO thin film SAW filters for a color television (15). The SAW devices act as a solid state band pass filter, which cannot be replaced by a Si-integrated circuit, and are composed of a layered structure of ZnO thin piezoelectric film on a glass substrate. The high quality growth techniques available for ZnO thin films have made possible the large scale production of these devices. This type of thin film device is used in a higher frequency region of GHz band for CATV and satellite TV.

Silicon Carbide (SiC) thin film high temperature sensors (16) are another attractive thin film device produced in the 1980's. They suggest the possibility of high accuracy, low temperature synthesis of high melting point materials by thin film growth processes.

Magnetic heads having a narrow magnetic gap for video tape recording systems and for computer disk applications are produced by thin film processing. In the production of the magnetic gap, a non-magnetic spacer has been formed from glass material. Prior to the use of thin film technology, the spacer manufacturing process was quite complex. For instance, magnetic head core material is first immersed in a mixed solution of finely-crushed glass, then taken out and subjected to centrifugation so that a homogeneous glass layer is deposited on the opposing gap surfaces of the core members. After forming a glass film on the core surfaces by firing the deposited glass layer, the two opposing gap faces are butted against each other with the glass layer sandwiched in between and then fused together by a heat treatment to form the desired operative gap. Since the width of the magnetic gap is around 0.3μ, these methods are difficult to use in production because of the difficulty in controlling the film thickness of the fired glass.

Thin film deposition technology overcomes these problems and realizes the production of the magnetic head with the narrow gap length of 0.3μ (17).

At present various kinds of thin film materials are used for the production of the electronic devices including high precision resistors, SAW filters, optical disks, magnetic tapes, sensors, and active matrix for liquid crystal TV (18). Recent progress of these thin film devices is the result of developments of Si-Large Scale Integration (LSI) technology including thin film growth process, microfabrication, and analysis technology of both the surface and interfaces of the thin films. Figure 1.2 shows photographs of these thin film devices.

(a) ZnO thin film SAW devices.

(b) PLZT, $(Pb, La)(Zr, Ti)O_3$ thin film optical waveguide switch.

Figure 1.2: Thin film electronic devices prepared by cathodic sputtering.

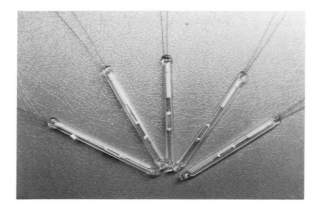

(c) SiC thin film high temperature sensors.

(d) Magnetic heads for VTR.

Figure 1.2: (continued) Thin film electronic devices prepared by cathodic sputtering techniques.

1.3 REFERENCES

1. Maissel, L.I. and Glang, R. (eds.) Handbook of Thin Film Technology New York, McGraw Hill (1970). Chopra, K.L., Thin Film Phenomena New York, McGraw Hill (1969). Vossen, J.L., and Kern, W., (eds.) Thin Film Processes New York, Academic Press (1978).

2. Read, M.H., and Altman, C. Appl. Phys. Lett., 7: 51 (1965).

3. Kitabatake, M., and Wasa, K., J. Appl. Phys., 58: 1693 (1985).

4. Wasa, K., and Kitabatake, M., in Thin Film Processing and Characterization of High-Temperature Superconductors, Series 3, (J.M.E.Harper, R.J.Colton, L.C.Feldman ed.). American Vac. Society, New York (1988).

5. Toxen, A.M., Phys. Rev., 123: 442 (1961), 124: 1018 (1961).

6. Hayakawa, S., and Wasa, K., Thin Film Materials Science and Technology Kyoritsu, Tokyo (1982).

7. Weimer, P.K., Proc. IRE, 50: 1462 (1962).

8. Esaki, L. Proc. 6th Int. Vac. Congr., Kyoto, (1974), Jpn. J. Appl. Phys., Suppl. 2, Pt. 1, 13: 821 (1974).

9. Kino, G.S. and Wagers, R.S. J. Appl. Phys., 44: 1480 (1973).

10. Miller, S.E., BSTJ, 48: 2059 (1969); Tien, P.K. Appl. Opt., 1O: 2395 (1971).

11. Spear, W.E., LeComber, P.G., J. Non-Cryst. Solids, 11: 219 (1972).

12. Spear, W.E., and LeComber, P.G., Solid State Commun., 17: 1193 (1975).

13. Carlson, D.E., Wronski, C.R., RCA Review, 38: 211 (1977).

14. Hamakawa, Y., Okamoto, H., Takakura, H. 18th IEEE Photovol. Spec. Conf., Las Vegas (1985).

15. Yamazaki, O., Mitsuyu, T., Wasa, K., IEEE Trans. Sonics and Ultrason., SU-27: 369 (1980).

16. Wasa, K., Tohda, T., Kasahara, Y., Hayakawa, S., Rev.Sci. Instr., 50: 1086 (1979).

17. K.Wasa, U.S. Patent 4,288,307 Sept. 1981, assigned to Matsushita Electric Corp.

18. Wasa, K., IECE J. Japan, 66: 707 (1983).

2

THIN FILM PROCESSES

Several publications have presented a detailed review of thin film deposition processes (1). Thus only brief descriptions of the thin film growth and deposition processes are presented in this chapter.

2.1 THIN FILM GROWTH PROCESS

Any thin film deposition process involves three main steps: (1) production of the appropriate atomic, molecular, or ionic species, (2) transport of these species to the substrate through a medium, and (3) condensation on the substrate, either directly or via a chemical and/or electrochemical reaction, to form a solid deposit. Formation of a thin film takes place via nucleation and growth processes. The general picture of the step-by-step growth process emerging out of the various experimental and theoretical studies can be presented as follows :

1. The unit species, on impacting the substrate, lose their velocity component normal to the substrate (provided the incident energy is not too high) and are physically adsorbed on the substrate surface.

2. The adsorbed species are not in thermal equilibrium with the substrate initially and move over the substrate surface. In this process they interact among themselves, forming bigger clusters.

3. The clusters or the nuclei, as they are called, are thermodynamically unstable and may tend to desorb in time depending on the deposition parameters. If the deposition parameters are such that a cluster collides with other adsorbed species before getting desorbed, it starts growing in size. After reaching a certain critical size, the cluster becomes thermodynamically stable and the nucleation barrier is said to have been overcome. This step involving the formation of stable, chemisorbed, critical-sized nuclei is called the nucleation stage.

4. The critical nuclei grow in number as well as in size until a saturation nucleation density is reached. The nucleation density and the average nucleus size depend on a number of parameters such as the energy of the impinging

species, the rate of impingement, the activation energies of adsorption, desorption, thermal diffusion, the temperature, topography, and chemical nature of the substrate. A nucleus can grow both parallel to the substrate by surface diffusion of the adsorbed species, as well as perpendicular to it by direct impingement of the incident species. In general, however, the rate of lateral growth at this stage is much higher than the perpendicular growth. The grown nuclei are called islands.

5. The next stage in the process of film formation is the coalescence stage, in which the small islands start coalescing with each other in an attempt to reduce the surface area. This tendency to form bigger islands is termed agglomeration and is enhanced by increasing the surface mobility of the adsorbed species, as, for example, by increasing the substrate temperature. In some cases, formation of new nuclei may occur on areas freshly exposed as a consequence of coalescence.

6. Larger islands grow together, leaving channels and holes of uncovered substrate. The structure of the films at this stage changes from discontinuous island type to porous network type. A completely continuous film is formed by filling of the channels and holes.

The growth process thus may be summarized as consisting of a statistical process of nucleation, surface-diffusion controlled growth of the three dimensional nuclei, and formation of a network structure and its subsequent filling to give a continuous film. Depending on the thermodynamic parameters of the deposit and the substrate surface, the initial nucleation and growth stages may be described as of (a) island type (called Volmer-Weber type), (b) layer type (called Frank-van der Merwetype), and (c) mixed type (called Stranski-Krastanov type). This is illustrated in Fig. 2.1. In almost all practical cases, the growth takes place by island formation. The subsequent growth stages for an Au film sputter deposited on NaCl at 25°C as observed in the electron microscope, are shown in Fig. 2.2.

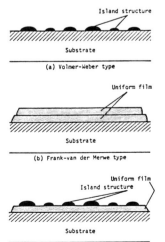

Figure 2.1: Three modes of thin film growth process.

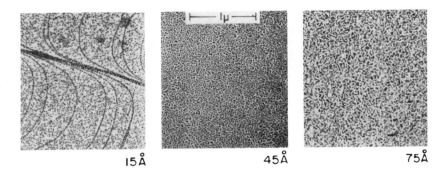

|15Å|45Å|75Å|

Figure 2.2: Transmission electron micrographs of 15, 45 and 75 Å thick argon sputtered Au films deposited on NaCl at 25°C at a deposition rate of approx.1Å/ sec.

Except under special conditions, the crystallographic orientations and the topographical details of different islands are randomly distributed, so that when they touch each other during growth, grain boundaries and various point and line defects are incorporated into the film due to mismatch of geometrical configurations and crystallographic orientations, as shown in Fig. 2.3. If the grains are randomly oriented, the films show a ring-type diffraction pattern and are said to be polycrystalline. However, if the grain size is small (20Å), the films show halo-type diffraction patterns similar to that exhibited by highly disordered or amorphous (noncrystalline) structures. It is to be noted that even if the orientation of different islands is the same throughout, as obtained under special deposition conditions (discussed later) on suitable single-crystal substrates, a single-crystal film is not obtained. Instead, the film consists of single-crystal grains oriented parallel to each other and connected by low angle grain boundaries. These films show diffraction patterns similar to those of single crystals and are called epitaxial/single crystal films.

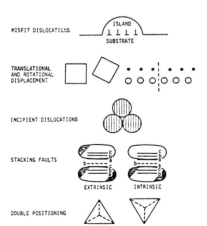

Figure 2.3: A schematic diagram showing incorporation of defects in a thin film during growth.

Besides grain boundaries, epitaxial films may also contain other structural defects such as dislocation lines, stacking faults, microtwins and twin boundaries, multiple-positioning boundaries, and minor defects arising from aggregation of point defects (for example: dislocation loops, stacking faults, and tetrahedra and small dotlike defects). Note that defects such as stacking faults and twin boundaries occur much less frequently in polycrystalline films. Dislocations with a density of 10^{10} to 10^{11} lines/cm^2 are the most frequently encountered defects in polycrystalline films and are largely incorporated during the network and hole stages, due to displacement (or orientation) misfits between different islands. Some other mechanisms which may give rise to dislocations in thin films are: (1) substrate film lattice misfit, (2) the presence of inherent large stresses in thin films, and (3) continuation of the dislocations ending on the substrate surface into the film.

After a continuous film is formed, the anisotropic growth takes place normal to the substrate in the form of cylindrical columns. The lateral grain size (or the crystallite size) of a film is primarily determined by the initial nucleation density. If, however, recrystallization takes place during the coalescence stage, the lateral grain size is larger than the average separation of the initial nuclei, and the average number of grains per unit area of the film is less than the initial nucleation density. The grain size normal to the substrates essentially equal to the film thickness for small $< 1\mu$ thicknesses. For thicker films, renucleation takes place at the surface of previously grown grains, and each vertical column grows multigranularly with possible deviations from normal growth.

2.1.1 Structural Consequences of the Growth Process

The microstructural and topographical details of a thin film of a given material depend on the kinetics of growth and hence on the substrate temperature, the source and energy of impurity species, the chemical nature, the topography of the substrate and gas ambients. These parameters influence the surface mobility of the adsorbed species: kinetic energy of the incident species, deposition rate, supersaturation (i.e.,the value of the vapor pressure/solution concentration above that required for condensation into the solid phase under thermodynamical equilibrium conditions), the condensation or sticking coefficient (i.e., the fraction of the total impinging species adsorbed on the substrate), and the level of impurities. Let us now see how the physical structure is affected by these parameters.

2.1.1.1 Microstructure: The lateral grain size is expected to increase with decreasing supersaturation and increasing surface mobility of the adsorbed species. As a result, deposits with well-defined large grains are formed at high substrate and source temperatures, both of which result in high surface mobility. Transmission electron micrographs of 100Å thick Au films deposited on NaCl at 100, 200, and 300°C by vacuum evaporation illustrate (Fig. 2.4) the effect of substrate temperature. Note that increasing the kinetic energy of the incident species (for example, by increasing the source temperature in the case of deposition by vacuum evaporation, or by increasing the sputtering voltage in the case of deposition by sputtering) also increases the surface mobility. However, at sufficiently high kinetic energies, the surface mobility is reduced due to the penetration of the incident species into the substrate, resulting in a smaller grain size. This effect of the kinetic energy of the impinging species on grain size is more pronounced at high substrate temperatures. Also, the effect of substrate temperature on grain size is more prominent for relatively thicker films.

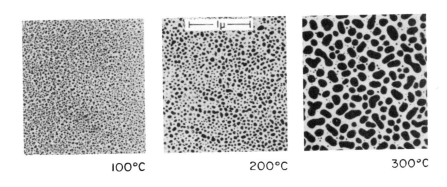

100°C   200°C   300°C

Figure 2.4: Transmission electron micrographs of 100 Å thick Au films vacuum evaporated on NaCl at 100, 200 and 300 °C.

The grain size may also be modified by giving the film a postdeposition annealing treatment at temperatures higher than the deposition temperature. The higher the annealing temperature, the larger is the grain size obtained. The effect of heat treatment is again more pronounced for relatively thicker films. It should be noted that the grain growth obtained during postdeposition annealing is significantly reduced from that obtained by depositing the film at annealing temperatures, because of the involvement of high-activation-energy process of thermal diffusion of the condensate atoms in the former case as compared to the process of condensation of mobile species in the latter.

For a given material-substrate combination and under a given set of deposition conditions, the grain size of the film increases as its thickness increases. However, beyond a certain thickness, the grain size remains constant, suggesting that coherent growth with the underlying grains does not go on forever and fresh grains are nucleated on top of the old ones above this thickness. This effect of increasing grain size with thickness is more prominent at high substrate temperatures. The effect of various deposition parameters on the grain size is summarized qualitatively in Fig. 2.5. It is clear that the grain size cannot be increased indefinitely because of the limitation on the surface mobility of the adsorbed species.

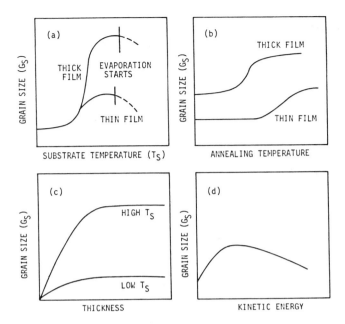

Figure 2.5: Qualitative representation of the influence of various deposition parameters on the grain size of thin films.

The formation of large-grain-sized epitaxial/single-crystal films under certain conditions has been mentioned earlier. The conditions favoring epitaxial growth are: high surface mobility as obtained at high substrate temperatures; low supersaturation; clean, smooth, and inert substrate surfaces; and crystallographic compatibility between the substrate and the deposit material. Films in which only a particular crystallographic axis is oriented along a fixed direction (due to preferential growth rate) are called oriented films. In contrast to epitaxial films which require a suitable single-crystal substrate, oriented films may also be formed on amorphous substrates. On the other extreme of thin film microstructures, highly disordered, very fine-grained, noncrystalline deposits with grain size 20Å and showing halo-type diffraction patterns similar to those of amorphous structures (i.e., having no translational periodicity over several interatomic spacings) are obtained under conditions of high supersaturation and low surface mobility. The surface mobility of the adsorbed species may be inhibited, for example, by decreasing the substrate temperature, by introducing reactive impurities into the film during growth, or by codeposition of materials of different atomic sizes and low surface mobilities. Under these conditions, the film is amorphous-like and grows layer-by-layer.

2.1.1.2 Surface Roughness: Under conditions of a low nucleation barrier and high supersaturation, the initial nucleation density is high and the size of the critical nucleus is small. This results in fine-grained, smooth deposits which become continuous at small thicknesses. On the other hand, when the nucleation barrier is large and the supersaturation is low, large but few nuclei are formed as a result of which coarse-grained rough films, which become continuous at relatively large thicknesses, are obtained. High surface mobility, in general, increases the surface smoothness of the films by filling in the concavities. One exception is the special case where the deposited material has a tendency to grow preferentially along certain crystal faces because of either large anisotropy in the surface energy or the presence of faceted roughness on the substrate.

A further enhancement in surface roughness occurs if the impinging species are incident at oblique angles instead of falling normally on the substrate. This occurs largely due to the shadowing effect of the neighboring columns oriented toward the direction of the incident species. Figure 2.6 shows the topography of two rough film surfaces, one (a) obtained by oblique deposition and the other (b) obtained by etching of a columnar structure. Also shown in the figure is the topography of a smooth and a rough CdS film prepared by controlled homogeneous precipitation (3) under different conditions.

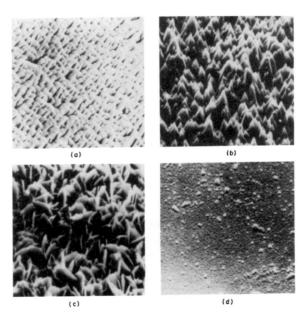

Figure 2.6: Scanning electron micrographs showing topography of smooth and rough films (a) obliquely deposited GeSe film, (b) etched CdS film (vacuum evaporated); (c) rough CdS film (solution grown); and (d) smooth CdS film (solution grown).

A quantitative measure of roughness, the roughness factor, is the ratio of the real effective area to the geometrical area. The variation of the roughness factor with thickness for a number of cases is qualitatively illustrated in Fig. 2.7. In the case of porous films, the effective surface area can be hundreds of times the geometrical area.

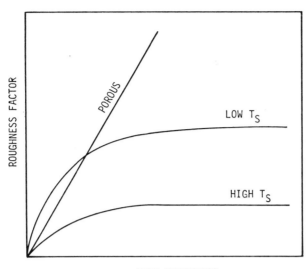

FILM THICKNESS

Figure 2.7: Qualitative variation of the roughness factor as a function of film thickness.

2.1.1.3 Density: Density is an important parameters of physical structure. It must be known for the determination of the film thickness by gravimetric methods. A general behavior observed in thin films is a decrease in the density with decreasing film thickness. This is qualitatively illustrated in Fig. 2.8.

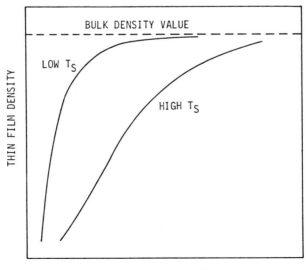

FILM THICKNESS

Figure 2.8: Qualitative variation of density as a function of film thickness.

Discrepancies observed in the value of the thickness at which the density of a given film approaches its bulk value are attributed to differences in the deposition conditions and measurement techniques employed by different observers.

In the case of porous films, which are formed due to incorporation of gaseous impurities under conditions of poor vacuum and high supersaturation, the density can be as low as 2 to 3% of the bulk density, even in thick films.

2.1.1.4 Adhesion: The adhesion of a film to the substrate is strongly dependent on the chemical nature, cleanliness, and the microscopic topography of the substrate surface. The adhesion of the films is better for higher values of (1) kinetic energy of the incident species, (2) adsorption energy of the deposit, and (3) initial nucleation density. The presence of contaminants on the substrate surface may increase or decrease the adhesion depending on whether the adsorption energy is increased or decreased, respectively. Also the adhesion of a film can be improved by providing more nucleation centers on the substrate, as by using a fine-grained substrate or a substrate precoated with suitable materials. Loose and porous deposits formed under conditions of high supersaturation and poor vacuum are less adherent than the compact deposits.

2.1.1.5 Metastable Structure: In general, departures from bulk values of lattice constants are found only in ultra-thin films. The lattice constants may increase or decrease, depending on whether the surface energy is negative or positive respectively. As the thickness of the film increases, the lattice constants approach the corresponding bulk values.

A large number of materials when prepared in thin film form exhibit new metastable structures not found in the corresponding bulk materials. These new structures may be purely due either to deposition conditions or they may be impurity/substrate stabilized. Some general observations regarding these new structures in thin films are as follows. (1) Most of the materials, in pure form or in combination with appropriate impurities, can be prepared in amorphous form. (2) The distorted NaCl-structure bulk materials tend to transform to the undistorted form in thin films. (3) The wurtzite compounds can be prepared in sphalerite form and vice versa. (4) Body centered cubic (bcc) and hexagonal close packed (hcp) structures have a tendency to transform to face centered cubic (fcc) structure. Some common examples of such abnormal structures found in thin films are: amorphous Si, Ge, Se, Te, and As; fcc Mo, Ta, W, Co, and β-Ta, etc. (all due to deposition conditions) ; and fcc Cr/Ni, bcc Fe/Cu, and fcc Co/Cu (all due to the influence of the substrate). Note that these abnormal metastable structures transform to the stable normal structures on annealing. This is illustrated in Fig. 2.9 for the case of ion-beam-sputtered Zr films.

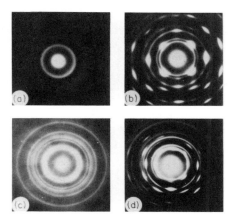

Figure 2.9 Electron diffraction patterns of $\simeq 500\mathring{A}$ thick Zr films, ion-beam sputtered onto NaCl at (a) 23°C, amorphous; (b) 250°C, fcc; (c) 450°C, hcp; and (d) fcc annealed at 675°C in vacuum, fcc + hcp.

2.1.2 Solubility Relaxation

Another consequence of the thin film growth process is the phenomenon of solubility relaxation. The atomistic process of growth during codeposition allows doping and alloying of films. Since thin films are formed from individual atomic, molecular, or ionic species which have no solubility restrictions in the vapor phase, the solubility conditions between different materials on codeposition are considerably relaxed. This allows the preparation of multicomponent materials, such as alloys and compounds, over an extended range of compositions as compared to the corresponding bulk materials. It is thus possible to have tailor-made materials with desired properties, which adds a new and exciting dimension to materials technology. An important example of this technology of tailor-made materials is the formation of hydrogenated amorphous Si films for use in solar cells. Hydrogenation has made it possible to vary the optical band gap of amorphous Si from 1 eV to about 2 eV and to decrease the density of dangling bond states in the band gap so that doping (n and p) is made possible.

2.2 THIN FILM DEPOSITION PROCESS

2.2.1 Classification of Deposition Process

Typical deposition methods of thin films are shown in Fig. 2.10. The deposition methods are composed of the physical vapor deposition (PVD) process and the chemical vapor deposition (CVD) process.

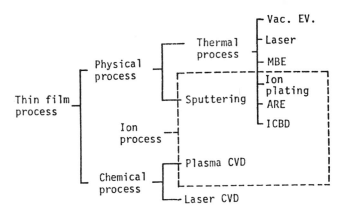

Figure 2.10: Thin film deposition processes.

PVD Process; The PVD process is divided into two categories; (1) thermal evaporation and (2) sputtering.

Thermal evaporation: Thermal evaporation process comprises evaporating source materials in a vacuum chamber below 1×10^{-6} Torr (1.3×10^{-4} Pa) and condensing the evaporated particles on a substrate. We conventionally call the thermal evaporation process as "vacuum deposition". Several types of the thermal evaporation process are proposed as shown in Fig. 2.11.

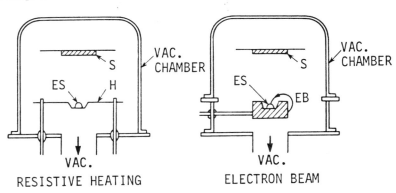

Figure 2.11: Thermal evaporation process: ES, evaporation source; S, substrate; H, heater; EB, electron beam source.

Resistive heating is most commonly used for the deposition of thin films. The source materials are evaporated by the resistively heated filament or boat, generally made of refractory metals such as W, Mo, and Ta, with or without ceramic coatings. Crucibles of quartz, graphite, alumina, beryllia, boron-nitride, and zirconia are used with indirect heating. The refractory metals are evaporated by electron beam deposition since source materials having a high melting point can not be evaporated by simple resistive heating.

Laser deposition is used for deposition of alloys and/or compounds with the controlled chemical composition. In laser deposition the high power pulsed laser, such as a KrF excimer laser (1 J/shot), is irradiated through a quartz window. A quartz lens is used to increase the energy density of the laser power on the target source. Atoms that are ablated or evaporated from the surface are collected on nearby sample surfaces to form thin films. Molecular beam epitaxy (MBE) process is the most reliable deposition process in thermal evaporation. Figure 2.12 shows a typical MBE system. The system is a controlled MBE process, where the evaporation rate of the source materials is controlled by in situ by a computerized process control unit. The man-made superlattice structure composed of thin alternating layers of GaAs and GaAlAs can be successfully deposited as shown in Fig. 2.13 (4).

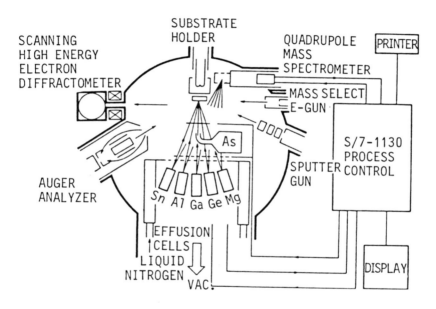

Figure 2.12: Molecular Beam Epitaxy (MBE) system (Esaki,(4))

This kind of deposition process is now widely used for the controlled deposition of alloys and compounds. The system is generally composed of a growth chamber, the analysis chamber, and sample chamber. Recently a vapor source was used for the MBE system as shown in Fig. 2.14, and metal-organic compounds are used for its source (5).

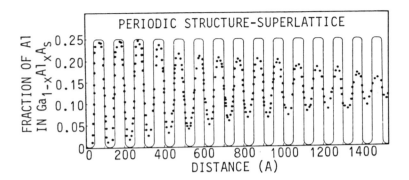

Figure 2.13: Man-made superlattice (Esaki,(4)).

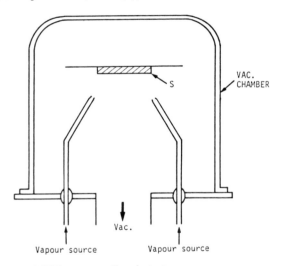

Figure 2.14: Vapor source MBE system: S, substrate.

Sputtering: When a solid surface is bombarded with energetic particles such as acceler-ated ions, surface atoms of the solid are scattered backward due to collisions between the surface atoms and the energetic particles as shown in Fig. 2.15. This phenomena is called "back-sputtering" or simply "sputtering". When a thin foil is bombarded with energetic particles some of the scattered atoms transmit through the foil. The phenomena is called "transmission-sputtering".

The word "spluttering" is synonymous with "sputtering". "Cathode sputtering", "Cathode disintegration" and "impact evaporation" are also used in the same sense.

Several sputtering systems are proposed for thin film deposition (6). Their designs are shown in Fig. 2.16.

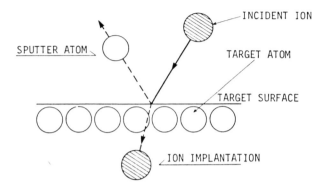

Figure 2.15: Physical sputtering processes.

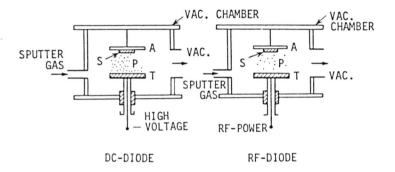

Figure 2.16: Sputter deposition system: A, anode; T, target; S, substrate; P, plasma.

Among these sputtering systems the simplest model is the dc diode sputtering system. The dc sputtering system is composed of a pair of planar electrodes. One of the electrodes is cold cathode and the other is anode. The front surface of the cathode is covered with target materials to be deposited. The substrates are placed on the anode. The sputtering chamber is filled with sputtering gas, typically Argon gas at 0.1 Torr. The glow discharge is maintained under the application of dc voltage between the electrodes. The Ar^+ ions generated in the glow discharge are accelerated at the cathode fall (sheath) and sputter the target resulting in the deposition of the thin films on the substrates. In the dc sputtering system the target is composed of metal, since the glow discharge (i.e. current flow) is maintained between the metallic electrodes.

By simple substitution of an insulator for the metal target in the dc sputtering discharge system, the sputtering discharge can not be sustained because of the immediate build-up of a surface charge of positive ions on the front side of the insulator. To sustain the glow discharge with the insulator target, an rf- voltage is supplied to the target. This system is called rf- diode sputtering. In the rf-sputtering system, the thin films of the insulator are sputtered directly from the insulator target.

When a reactive gas species such as oxygen or nitrogen is introduced into the chamber, thin films of compounds (i.e. oxides or nitrides) may be deposited when sputtering the appropriate metal targets. This technique is known as reactive sputtering, and may be used in either the dc or rf mode.

In magnetron sputtering a magnetic field is superposed on the cathode and glow discharge which is parallel to the cathode surface. The electrons in the glow discharge shows cycloidal motion and the center of the orbit drifts in a direction of ExB with the drift velocity of E/B, where E and B denote the electric field in the discharge and the superposed transverse magnetic field, respectively. The magnetic field is oriented such that these drift paths for electrons form a closed loop. This electron trapping effect increases the collision rate between the electrons and the sputtering gas molecules. This enables one to lower the sputtering gas pressure as low as 10^{-4} Torr, but more typically 10 mTorr. In the magnetron sputtering system, the magnetic field increases the plasma density which leads to increases of the current density at the cathode target, effectively increasing the sputtering rate at the target. Due to the low working gas pressure, the sputtered particles traverse the discharge space without collisions, which results in effectively a higher deposition rate than higher pressure deposition systems.

At present the planar magnetron is indispensable for the fabrication of semiconductive devices. Historically, magnetron sputtering was first proposed by Penning in 1936 (7). A prototype of the planar magnetron was invented by Wasa in 1967 (8), and Chapin improved this system (9). A typical construction is shown in Fig. 2.17.

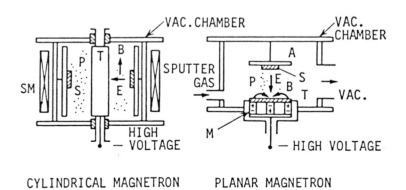

CYLINDRICAL MAGNETRON   PLANAR MAGNETRON

Figure 2.17: Magnetron sputtering system: A, anode; T, target; P, plasma; SM, solenoid magnet; M, magnet; E, electric field; B, magnetic field.

In these glow discharge systems the sputtered films are irradiated by sputtering gas molecules during thin film growth. This causes the inclusion of the gas molecules in the sputtered films. In the ion beam sputtering system, incident ions are generated at the ion source. The target is sputtered in a sputtering chamber separated from the ion source. The typical ion beam current is 10 to 500 mA, with an ion energy from is 0.5 to 2.5 kV. Since the ions are generated in the ion source discharge chamber, the working pressure of the sputtering chamber can be reduced as low as $1x10^{-5}$ Torr. This reduces the amount of gas molecules included in the sputtered films.

Pioneering work was done by Chopra on the deposition of thin films by ion beam sputtering in 1967 (10). Although ion beam sputtering is not widely used for thin film deposition, this kind of system is widely used for the sputter etching of semiconductive devices (11). Recent interest has been paid to the synthesis of exotic thin films by ion beam sputtering (12). The basic sputtering data has been summarized by Behrish (13).

Miscellaneous PVD processes: Ion plating was first proposed by Mattox in the 1960's (14). The coating flux is usually provided by thermal evaporation. The evaporated atoms are ionized at the plasma region and accelerated by the electric field prior to deposition. A typical construction is shown in Fig. 2.18. The adhesion of thin films is improved by the acceleration of evaporated atoms.

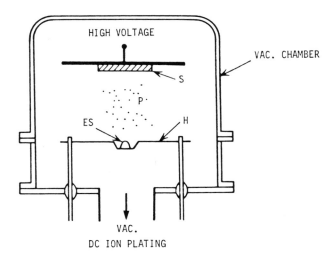

Figure 2.18: Ion plating: S, substrate; ES, evaporation source; P. plasma; H, heater.

Activated Reactive Evaporation (ARE) proposed by Bunshah is commonly used for the deposition of metal oxides, carbides and nitrides (15). The configuration of the ARE system shown in Fig. 2.19 is similar to the ion plating system. Reactive gas is injected into the plasma region so as to achieve the reaction between evaporated atoms and the reactive gas atoms.

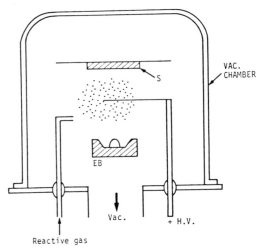

Figure 2.19: Activated Reactive Evaporation (ARE): S, substrate; EB, electron beam evaporation source.

Ionized cluster beam deposition (ICBD), which was developed by Takagi in the 1970's is a modification of ion plating (16). The construction is shown in Fig. 2.20. Atoms are evaporated from a closed source through a nozzle. Cooling of the atoms upon expansion through the nozzle leads to cluster formation, which might have a few hundred to 1000 atoms per cluster. The cluster is ionized through the plasma region and then is accelerated to the substrate. The average energy of the atoms in the accelerated are in a range from 0.2 to several eV, even when the clusters themselves are accelerated to kilovolts. The relatively low energy of the adatoms will reduce the lattice damage to the substrate surface.

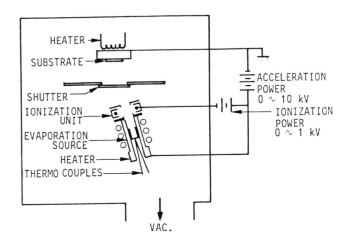

Figure 2.20: Ionized Cluster Beam (ICB) deposition system (Takagi, 1975 (16)).

CVD Processes: When a volatile compound of the substance to be deposited is vaporized, and the vapor is thermally decomposed or reacted with other gases, vapors, or liquids at the substrate to yield nonvolatile reaction products which deposit atomistically on the substrate, the process is called chemical vapor deposition.

Most CVD processes operate in the range of a few Torr to above atmospheric pressure of the reactants. A relatively high temperature (near 1000°C) is required for CVD processes. Several CVD processes are proposed to increase the efficiency of the chemical reaction at lower substrate temperature. Typical construction of the CVD deposition system is shown in Fig. 2.21.

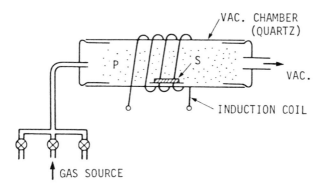

Figure 2.21: Chemical Vapor Deposition (CVD):   S, substrate.

Plasma-assisted chemical vapor deposition (PACVD) is one of the modifications of conventional CVD. The typical construction is shown in Fig. 2.22. In the PACVD system, the electric power is supplied to the reactor so as to generate the plasma. Usually the working pressure is in the range of 0.1 to 1.0 Torr. In the plasma, the degree of ionization is typically only 10^{-4}, so the gas in the reactor consists mostly of neutrals. Ions and electrons will travel through the neutrals and get energy from the electric field in the plasma. The average electron energy is 2 to 8 eV, which corresponds to the electron temperature of 23,000 to 92,800 K. In contrast, the heavy, much more immobile ions cannot get effectively couple energy from the electric field. The ions in the plasma show slightly higher energy than the neutral gas molecules of a room temperature. Typically the temperature of the ions in a processing plasma is around 500K.

Since the electron temperature in the plasma is much higher than the gas temperature, thermal equilibrium is not maintained between electrons and neutral gas molecules. This suggests that the plasma in the glow discharge is a sort of "cold plasma" which comprises high temperature electrons, i.e. "hot electrons" and room temperature gas molecules. The high temperature electrons enhance the chemical reactions in the plasma as indicated in Fig. 2.23. This results in the lowering the temperature of reactions. For this reason, PACVD is one of the most important processes in the electronics industry.

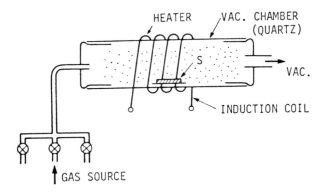

Figure 2.22: Plasma Assisted Chemical Vapor Deposition (PACVD): S, substrate; P, plasma.

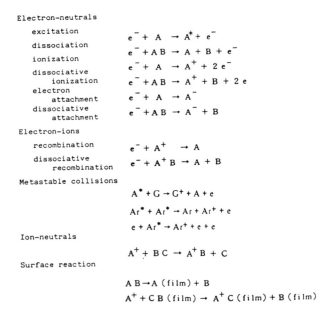

Figure 2.23: Plasma enhanced chemical reactions.

Several improved PACVD processes have recently been developed. In one major development, microwaves-based plasmas have been used to reduce the working pressure. A magnetic field is superposed on the microwave plasma at the appropriate field strength to cause a resonance between the electron cyclotron frequency and the applied electric field. This in known as an Electron Cyclotron Resonance (ECR) condition. A typical construction is shown in Fig. 2.24.

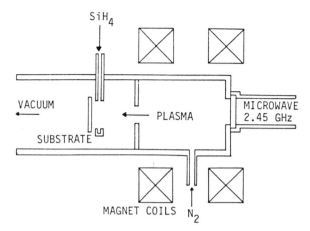

Figure 2.24 ECR plasma assisted chemical vapor deposition.

Laser-assisted chemical vapor deposition (laser CVD) has also been recently developed as a modification of CVD (17). The CVD reactions are activated by the irradiation of the ultra-violet laser light without the application of the electric power.

2.2.2 Deposition Conditions

Most thin films are deposited by evaporation (vacuum deposition), sputtering and/or chemical vapor deposition. Table 2.1 shows a guide for the deposition of the elements. A suitable selection of deposition processes are required when thin films are used for the preparation of the active electronic devices such as semiconducting devices, since the surface of the semiconductive substrates is often damaged during deposition. The special features of the typical deposition processes are listed in Table 2.2.

The nature of deposited films is governed by the deposition parameters including deposition rate, substrate temperature, substrate materials, and deposition atmosphere (18).

The chemical composition of deposited films is governed by the substrate temperature and/or the deposition atmosphere. Under low substrate temperatures the chemical composition of deposited films coincides to that of the source materials. Under high substrate temperatures the chemical composition of deposited films differs from the source materials due to the reevaporation of high vapor pressure materials from the films during the deposition.

The gas molecules of the deposition atmosphere are frequently included in the deposited films. The chemically active gas molecules react with the thin films during film growth and the resultant films become the compounds between the evaporated source and the active gas in the deposition atmosphere. Thin films of metal oxides, for instance, are prepared by the reactive sputtering from metal target in oxygen atmosphere.

Legend (key box):

ATOMIC NO.
ELEMENT
ATOMIC WEIGHT
DENSITY (g/cc)
ρ (μΩ cm)
MP (°C)
VP (°C) CRYSTAL
(S) TYPE
DEPOSITION METHODE

DEPOSITION METHODE

EB   ELECTRON BEAM
R( ) RESISTIVE HEAT(HEATER)
CR( ) RESISTIVE HEAT(CRUCIBLE)
BO( ) RESISTIVE HEAT(BOAT)
SP   SPUTTERING
MP   MELTING POINT
VP   TEMPERATURE AT VAPOUR PRESSURE, 1×10^{-4} TORR
S   SPUTTERING YIELD FOR 600eV Ar ION (Atoms/Ion)
ρ   RESISTIVITY

	I a	II a	III a	IV a	V a	VI a	VII a	VIII		
2	Li^3 6.939 0.534 8.55 181 *404 bcc EB	Be^4 9.0122 1.85 2.8 1284 *997 hcp								
3	Na^{11} 22.9898 0.97 4.2 97.6 *93 bcc EB	Mg^{12} 24.312 1.74 3.9 650 *327 hcp								
4	K^{19} 39.102 0.86 9.1 63.4 *123 bcc	Ca^{20} 40.08 1.55 3.43 839 *459 fcc(bcc)(hcp) CR(FQ)	Sc^{21} 44.956 3.00 66 1539 *1107 hcp EB.SP	Ti^{22} 47.90 4.51 50 1668 *1442 hcp (0.6)(bcc) EB.SP	V^{23} 50.942 6.10 18.2 1905 *1547 bcc EB.SP	Cr^{24} 51.996 7.19 12.7 1875 *1157 bcc (1.3) BO(W).SP	Mn^{25} 54.9380 7.43 258 1244 bcc *747 (fcc)(cub) EB.BO(W)	Fe^{26} 55.847 7.86 8.9 1535 bcc *1227 fcc EB.SP	Co^{27} 58.9332 8.90 5.6 1492 *1257 hcp (1.4)(fcc) EB.SP	
5	Rb^{37} 85.47 1.532 11.0 38.6 *94 bcc	Sr^{38} 87.62 2.6 23 772 *404 fcc EB.BO(W)	Y^{39} 88.905 4.47 53 1502 *1332 hcp EB.SP	Zr^{40} 91.22 6.49 40 1850 *1987 hcp(bcc) EB.SP	Nb^{41} 92.906 8.4 13.9 2468 *2277 (0.65) bcc EB.SP	Mo^{42} 95.94 10.22 5.2 2615 *2117 (0.9) bcc EB.SP	Tc^{43} 99 11.5 - 2170 *2077 EB	Ru^{44} 101.07 12.2 7.6 2280 *1987 hcp EB.SP	Rh^{45} 102.905 12.4 4.3 1960 *1707 (1.5) fcc EB.SP	
6	Cs^{55} 132.905 1.90 18.8 28.6 *78 bcc EB	Ba^{56} 137.34 3.5 60 725 *462 bcc EB	57~71 ACTINIDE	Hf^{72} 178.49 13.1 29.6 2222 *1997 bcc EB.SP	Ta^{73} 180.948 16.6 12.6 2998 *2587 (0.3) bcc EB.SP	W^{74} 183.85 19.3 4.9 3380 *2757 (0.6) bcc EB.SP	Re^{75} 186.2 21.02 18.6 3160 *2587 (0.9) bcc EB.SP	Os^{76} 190.2 22.5 9.66 3027 *2487 (0.95) hcp EB.SP	Ir^{77} 192.2 22.5 4.9 2443 *2107 fcc EB.SP	
7	Fr^{87} 223 - - 24 *61	Ra^{88} 226 5.0 - 700 *417	Ac^{89} 227 10.07 - 1050 *1332	Th^{90} 232.038 11.7 13 1751 *1977 (0.7) fcc	Pa^{91} 231 15.4 - 1425 * -	U^{92} 238.03 19.07 32 1131 ort *1582 (1.0) (tet)(bcc) EB.SP	Np^{93} 237 19.5 - 637 ort * - (tet)(bcc)	Pu^{94} 242 19-19.72 150(25°C) 640 mnc *1207 (ort)(bcc)(fcc.fct)	Am^{95} 243 11.7 - 995 *867	

LANTHANIDE 57~71

	III a	IV a	V a	VI a	VII a		VIII	
	La^{57} 138.91 6.17 57 920 *1422 hcp(bcc)(fcc) EB	Ce^{58} 140.12 6.67 75 797 *1377 fcc(hcp)(bcc) EB	Pr^{59} 140.907 6.77 68 935 *1147 hcp(bcc) EB	Nd^{60} 144.24 7.004 64 1024 *1047 hcp(bcc) EB	Pm^{61} 147 - - 1035 * - EB	Sm^{62} 150.35 7.54 92 1072 *580 rmb EB	Eu^{63} 151.96 5.26 81 826 *466 bcc EB	

Table 2.1: Periodic table with deposition conditions.

H¹ 1.00797 / 0.071 / − / −259.2 / •−	**METAL SEMICONDUCTOR** — **NON-METAL**	**He²** 4.0026 / − / − / − / •−

Legend:

- fcc Face-centered cubic
- bcc Body-centered cubic
- hex Hexagonal
- hcp Closed packed hexagonal
- dia Diamond type
- cub Cubic
- rmb Rhombohedral
- tet Tetragonal
- fct Face-centered tetragonal
- ort Orthorhombic
- mnc Monoclinic

	IIIb	IVb	Vb	VIb	VIIb
Period 2	**B⁵** 10.811 / 2.34 / 1.8×10^{18} / 2225 / •1707 / EB.SP	**C⁶** 12.01115 / 2.26 / (350~6300) / 3827 / •2137 dia hex / EB	**N⁷** 14.0067 / 0.81 / − / −210 / •− cub	**O⁸** 15.9994 / 1.14 / − / −218.8 / •− ort	**F⁹** 18.9984 / 1.5(−273°C) / − / −219.61 / •−
(VIIIb/Ne)					**Ne¹⁰** 20.183 / 1.20(−245°C) / − / −248.59 / •− fcc

	IIIb	IVb	Vb	VIb	VIIb	
Period 3	**Al¹³** 26.9815 / 2.70 / 2.45 / 660.1 / •972 / (1.2) fcc / EB.R(W).SP	**Si¹⁴** 28.086 / 2.33 / 3.5×10^{11} / 1412 / •1337 / (0.2) dia / EB	**P¹⁵** 30.9738 / 1.82 / − / 44.0 / •129 / ort	**S¹⁶** 32.064 / 2.07 / − / 119 / •55 / ort	**Cl¹⁷** 35.453 / 2.2(−273°C) / − / −100.99 / •− / ort / CR(FQ)	**Ar¹⁸** 39.948 / 1.65(−233°C) / − / −189.37 / •− / fcc

	Ib	IIb	IIIb	IVb	Vb	VIb	VIIb	
Period 4	**Ni²⁸** 58.71 / 8.90 / 6.14 / 1453 / •1262 / fcc / EB.SP							
	Cu²⁹ 63.546 / 8.96 / 1.56 / 1083 / •1027 / (2.3) fcc / EB.SP	**Zn³⁰** 65.37 / 7.14 / 5.5 / 419.505 / •247 / hcp / EB.SP	**Ga³¹** 69.72 / 5.91 / 13.6 / 29.6 / •907 / ort / CR(FQ)	**Ge³²** 72.59 / 5.32 / 4.0×10^{7} / 936 / •1137 / dia / EB	**As³³** 74.9216 / 5.72 / 32 / 817 / •204 / rmb / CR(Al₂O₃)	**Se³⁴** 78.96 / 4.79 / 8×10^{6} / 217 / •164 / hex / BO(W), EB	**Br³⁵** 79.904 / 4.2(−273°C) / − / −7.2 / •− / ort	**Kr³⁶** 83.80 / 3.4(−273°C) / − / −157.3 / •− / fcc

	Ib	IIb	IIIb	IVb	Vb	VIb	VIIb	
Period 5	**Pd⁴⁶** 106.4 / 12.02 / 10.0 / 1552 / •1192 / (2.4) fcc / R(W).EB.SP							
	Ag⁴⁷ 107.868 / 10.5 / 1.51 / 960.8 / •832 / (3.8) fcc / EB.R(W).SP	**Cd⁴⁸** 112.40 / 8.65 / 6.8 / 321.03 / •177 / hcp / CR(FQ)	**In⁴⁹** 114.82 / 7.31 / 8.4 / 156.61 / •742 / fct / EB. BO(W)	**Sn⁵⁰** 118.69 / 7.30 / β11.5 / 231.91 / •997 / dia (tet) / CR(Al₂O₃). SP	**Sb⁵¹** 121.75 / 6.62 / 39.0 / 630.5 / •425 / rmb / CR(Al₂O₃), SP	**Te⁵²** 127.60 / 6.24 / 4.2×10^{5} / 449.6 / •280 / hex / CR(FQ)	**I⁵³** 126.9044 / 4.94 / 13×10^{21} / 113.6 / •− / ort	**Xe⁵⁴** 131.30 / − / − / −112.5 / •− / fcc

	Ib	IIb	IIIb	IVb	Vb	VIb	VIIb	
Period 6	**Pt⁷⁸** 195.09 / 21.45 / 9.81 / 1769 / •1747 / (1.6) fcc / EB.SP							
	Au⁷⁹ 196.967 / 19.3 / 2.04 / 1063 / •1132 / (2.8) fcc / EB.R(W).SP	**Hg⁸⁰** 200.59 / 13.6 / 94.0766 / −38.87 / •−7 / rmb	**Tl⁸¹** 204.37 / 11.85 / 15 / 303 / •463 / hcp (bcc) / CR(FQ)	**Pb⁸²** 207.19 / 11.4 / 19.0 / 327.426 / •547 / fcc / EB, BO(W)	**Bi⁸³** 208.980 / 9.8 / 107 / 271.375 / •517 / rmb / R(W),CR(Al₂O₃)	**Po⁸⁴** 210 / 9.2 / − / 246 / •221 / mnc / CR(FQ)	**At⁸⁵** 210 / − / − / 302 / •43	**Rn⁸⁶** 222 / − / − / −71 / •−

Cm⁹⁶	Bk⁹⁷	Cf⁹⁸	Es⁹⁹	Fm¹⁰⁰	Md¹⁰¹	No¹⁰²	Lw¹⁰³	
247	249	251	254	253	256	−	−	
− / − / − / •−	− / − / − / •−	− / − / − / •−	− / − / − / •−	− / − / − / •−	− / − / − / •−	− / − / − / •−	− / − / − / •−	

Gd⁶⁴	Tb⁶⁵	Dy⁶⁶	Ho⁶⁷	Er⁶⁸	Tm⁶⁹	Yb⁷⁰	Lu⁷¹	
157.25 / 7.895 / 134 / 1312 / •1077 / (bcc) hcp / EB	158.924 / 8.272 / 116 / 1356 / •1147 / hcp / EB.SP	162.50 / 8.54 / 91 / 1407 / •897 / hcp / EB	164.930 / 8.803 / 94 / 1461 / •947 / hcp / EB	167.26 / 9.051 / 86 / 1497 / •947 / hcp / EB	168.934 / 9.332 / 90 / 1545 / •680 / hcp / CR(Al₂O₃)	173.04 / 6.96 / 28 / 824 / •417 / fcc (bcc)	174.97 / 9.842 / 68 / 1652 / •1277 / hcp	

Type of deposition	Evaporation		Sputtering		CVD	
Property	Resistive heating	Electron beam	Diode	Magnetron	Pyrolysis	Plasma
Thin film material	Material of low melting point	Material of high melting point, refractory metals	Wide varieties of materials, compounds refractory metals, alloys		Decomposition and/or chemical reaction of organometallic compounds or halides	
Substrate temperature	low		high (> 300 °C)	low (∿ 100 °C)	high (∿ 1000 °C)	high (> 300 °C)
Deposition rate	high metal; 0.5 ∿ 5 µm/min		low metal; 0.02 ∿ 0.2 µm/min	high same rate to evaporation	high same rate to evaporation	
Gas pressure	low < 10^{-5} Torr		high $10^{-2} \sim 10^{-1}$ Torr	low $10^{-4} \sim 10^{-3}$ Torr	high 1 at	high 1 ∿ 10 Torr
Energy of evaporated atoms	0.1 ∿ 0.2 eV		10 ∿ 200 eV	10 ∿ 20 eV		
Energy of adatoms	0.1 ∿ 0.2 eV		0.1 ∿ 20 eV	0.2 ∿ 10 eV		
Contamination Residual gas atom number / adatom number	∿ 10^{-2} [high purity in MBE system]		50	1	high purity	

Table 2.2: Special features of deposition processes.

The crystalline properties of the deposited films are controlled by the selection of the substrate materials and the substrate temperature. Amorphous thin films are prepared on a glass and/or ceramic substrate at substrate temperatures below the crystallization temperature of the thin films. Polycrystalline films are prepared on a glass and/or ceramic substrates at the substrate temperatures above the crystallization temperature. Single crystalline films are prepared on a single crystal substrate due to the epitaxial growth process and, in general, they are epitaxially grown on a single crystal substrate at the substrate temperature above the epitaxial temperature.

In the epitaxial growth process the epitaxial temperature is governed by the relationship

$$R = a \exp(-Q/kT_e), \qquad (2.1)$$

where R denotes the deposition rate; T_e, the minimum temperature for the epitaxial growth (epitaxial temperature); and Q the activation energy for the epitaxial growth. The epitaxial temperature is about 400°C for vacuum evaporated Ge thin films, and about 1,100 to 1,200°C for Si thin films. A low the deposition rate results in lower epitaxial temperature. These epitaxial processes are called vapor phase epitaxy. In contrast, polycrystalline films deposited on single crystalline substrate typically become single crystalline films due to a postannealing process. This case is called solid phase epitaxy. The relation between the deposition conditions and the crystalline properties of the deposited films are summarized in Fig. 2.25.

BASIC PROCESS

1. AMORPHOUS PHASE   Ts < Tc

2. POLYCRYSTALLINE   Ts > Tc

                            Ts < Tc , post-anneling

3. SINGLE CRYSTALS   Ts > Tepi.
   (single crystal sub.)   Ts < Tc, post-annealing
                            (solid-phase epitaxy)

Figure 2.25: Deposition conditions and the crystalline properties.

Typical substrate materials of glass, ceramics and single crystal are shown in Tables 2.3, 2.4, and 2.5, respectively. The epitaxial relations of the single crystal substrate are shown for various thin films in Table 2.6.

Glass property		Microsheet	Pyrex	Alumino borosilicate	Fused silica
Corning code No.		0211	7740	7059	7940
Density	(g/cm^3)	2.57	2.23	2.76	2.20
Thermal expansion ($\times 10^{-7}$/°C) (0 $\sim$ 300°C)		73.8	32.5	46	5.5
Strain point	(°C)	508	510	593	956
Anneal point	(°C)	550	560	639	1084
Softening point	(°C)	720	821	844	1580
Thermal conductivity (cal/cm/sec/°C) (25°C)		-	0.0027	-	0.0034
Hardness	(KHN$_{100}$)	608	418	-	489
Young's modulus ($\times 10^3$kg/cm^2)		7.59	6.4	6.89	7.4
Poisson's ratio		0.22	0.20	0.28	0.16
Resistivity (log ρ, Ωcm)	25°C	-	15	-	17
	250°C	8.3	8.1	13.1	11.8
Dielectric constant (1 MHz, 20°C)		6.7	4.6	5.84	3.8
tan δ	(%)	3.1	2.6	0.58	0.0038
Refractive index (5893, 5876Å)		1.523	1.474	1.530	1.459
Optical transparency		> 90% 0.36 $\sim$ 2.5µm (1 mm thick)	> 90% 0.36 $\sim$ 2.2µm (2 mm thick)	-	> 90% 0.36 $\sim$ 2.2µm (2 mm thick)
Composition		SiO_2 64.4% B_2O_3 10.3% Na_2O 6.2% ZnO 5.4% K_2O 6.9% Al_2O_3 4.1% TiO_2 3.1%	SiO_2 80.5% B_2O_3 12.8% Na_2O 4.1% Al_2O_3 2.5% K_2O 0.5%	SiO_2 49.9% BaO 25.1% B_2O_3 10.5% Al_2O_3 10.3% CaO 4.3%	SiO_2 99.5%

Table 2.3: Typical glass substrates for growth of polycrystalline thin films.

Property of ceramics	Steatite MgO·SiO$_2$	Forsterite 2 MgO· SiO$_2$	Alumina Al$_2$O$_3$ 96~97%	Alumina Al$_2$O$_3$ 99.5%	Beryllia BeO	Spinel MgO· Al$_2$O$_3$
Density (g/cm^3)	2.6~2.7	2.7~2.8	3.8	3.8~3.9	2.9	3.3
Thermal expansion (×10^{-6}/°C) (40~400°C)	7~8	10	6.7	6.8	6	8.1
Max. safe temp. (°C)	1000	1000	1600	1750	1600	1200
Thermal conductivity (cal/cm/sec/°C) (20°C)	0.006	0.008	0.05	0.06~0.07	0.5~0.55	0.04
Specific heat (cal/g·°C)	—	—	0.19	0.19	—	—
Hardness, Hv (kg/mm^2)	650 (500g)	900 (500g)	1600 (500g)	1600 (500g)	—	—
Compressive strength (kg/cm^2)	9000	9000	17,500	25,000	—	—
Flexural strength (kg/cm^2)	1650	1650	2800 ~3500	2800 ~4000	—	1500
Young's modulus (×10^{6}kg/cm^2)	—	—	3.4~3.5	~3.7	—	—
Dielectric strength (kV/mm)	9	9~10	10	10	9	10
Resistivity (log ρ, Ωcm) 25°C	>14	>14	>14	>14	>14	>14
Resistivity (log ρ, Ωcm) 500°C	8	10	11	11	—	11
Dielectric constant (1 MHz) (25°C)	6.3	6~6.5	9.4	9.7	6.3~6.4	8
tan δ(%) (1 MHz) (25°C)	0.06	0.01 ~0.03	0.02	0.02	0.01	0.01
Miscellaneous property			surface roughness $R_t = 1.3\,\mu$ $R_r = 0.5\,\mu$ $R_a = 0.3\,\mu$ $R_n = 0.075\mu$			

Table 2.4: Typical ceramic substrates for the growth of polycrystalline thin films.

Crystal property	α-Al₂O₃ (sapphire)	Si	Ge	GaAs	MgO	MgO·Al₂O₃ (spinel)	NaCl	Mica	SiO₂	LiNbO₃
Crystal system	trigonal	cubic	cubic	cubic	cubic	cubic	cubic	monoclinic	hexagonal	trigonal
Lattice constant (Å)	a=4.763 c=13.003	a=5.431	a=5.657	a=5.654	a=4.203	·	a=5.628			a=5.148 c=13.163
Density (g/cm³)	3.97	2.33	5.32	5.32	3.65	3.58~3.61	2.16	2.76~3.0	2.65	4.64
Thermal expansion (×10⁻⁷/°C)(25°C)	53 $(C_\parallel)$ 45 $(C_\perp)$	26	58	57	138	76		110 $(C_\parallel)$ 200 $(C_\perp)$	79.7 $(C_\parallel)$ 133.7 $(C_\perp)$	
Melting point (°C)	2053	1430	937	1238	2800	2135	801		1425	1253
Thermal conductivity (cal/cm/sec/°C)(25°C)	0.1	0.34	0.15	0.11	0.06	0.06	0.0155	0.0016	0.03	
Specific heat (cal/g/°C)	0.18	0.17	0.077	0.038	0.209	0.2	0.204	0.21	0.191	
Hardness, mohs	9 (H_k=2300 kg/mm²)	7	6.5		5.5	8	2.5	2.5~3.0	7	5
Compressive strength (kg/cm²)	30000									
Flexural strength (kg/cm²)	7000					2800~3500				
Young's modulus (×10⁶ kg/cm²)	4.8	1~1.6		0.6~1.2	2.5~3.5					
Dielectric strength (kV/mm)	48 (25°C)					>14 (25°C)				
Resistivity (log ρ, Ωcm)	16 (25°C) 11 (500°C)	5.8 (25°C)	1.66(25°C)						16(20°C)	
Dielectric constant (25°C)	11.5 $(C_\parallel)$ 9.3 $(C_\perp)$	11.7	13.8	12.9	9.65	8.4	5.62	6.5~9.0	4.6 $(C_\parallel)$ 4.5 $(C_\perp)$	
tan δ (%) (1 MHz)(25°C)	0.01					0.008				
Refractive index	N_0=1.768 N_e=1.760	N_0=3.47 (λ=1.66μ)	N_0=4.10 (λ=2.06μ)	N_0=3.36 (λ=1μ)	N_0=1.74 (λ=0.633μ)	N_0=1.718 (λ=0.5893μ)		N_0=1.544 (N~D)		N_0=2.297 (λ=0.63μ) N_e=2.208
Optical transparency	>80% 0.25~4 μm (1 mm thick)					>85% 0.3~5 μm				
Miscellaneous property	(0001), c plane (1102), R plane (1120), a plane							chemical formula $H_2Al_3(SiO_4)_3$		

Table 2.5: Typical crystal substrates for the epitaxial growth of thin films.

2.3 CHARACTERIZATION

A measurement of thin film properties is indispensable for the study of thin film materials and devices. The chemical composition, crystalline structure, optical properties, electrical properties, and mechanical properties must be considered in evaluating thin films.

Several methods are proposed for the evaluation of the thin films. Table 2.7 shows a summary of the methods used. Among these processes, a rapid progress has been made in the evaluation of the surface and thin film composition for semiconductor materials. Several methods have been proposed for the evaluation of the surface or thin films. Typical methods are listed in Table 2.8. There are a number of major considerations that determine the choice of an instrumental method to solve a specific problem in the surface or thin films including area and depth to be sampled, sensitivity and reproducibility, and the number of detectable elements.

In the daily study of thin film materials we should evaluate the thin film properties listed in Table 2.7 and grasp a correlation between a growth condition and the properties of the resultant thin films as shown in Fig. 2.26.

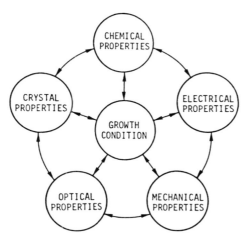

Figure 2.26: Correlations between growth conditions and the properties of the resultant thin films.

In-situ evaluations of the surface or thin film properties are needed for a determination of the relation between the growth condition and the film properties. Auger Electron Spectrometry (AES) is widely used for measurement of a chemical composition of the thin film during film growth. The condensation process and the surface crystallinity of the thin film are evaluated by RHEED.

Aside from instrumental analyses, the evaluation of electrical properties or physical properties are also important. For example, the contents of impurity in a metal film are evaluated by the measurements of an electrical resistance and its temperature coefficient (47).

Thin Film	Substrate	Epitaxial Relation	Ref.
Ag	Si	Ag(111)//Si(111)	19)
Au	NaCl	(100)Au//(100)NaCl, [110]Au//[110]NaCl	20)
Cu	Ti	(111)Cu//(0001)Ti, [1$\bar{1}$0]Cu//[1$\bar{1}$20]Ti	21)
	W	(111)Cu//(110)W, [$\bar{1}\bar{1}$2]Cu//[$\bar{1}$10]W.	
Co	MgO	(001)Co//(001)MgO, [100]Co//[100]MgO	22)
Fe	MgO	(001)Fe//(001)MgO, [100]Fe//[110]MgO	23)
Si	α-Al$_2$O$_3$	(100)Si//(1$\bar{1}$02)Al$_2$O$_3$, [110]Si//[$\bar{2}$201]Al$_2$O$_3$	24)
		(111)Si//(0001)Al$_2$O$_3$, [1$\bar{1}$0]Si//[$\bar{1}$230]Al$_2$O$_3$	
		(111)Si//(11$\bar{2}$0)Al$_2$O$_3$, [1$\bar{1}$0]Si//[$\bar{2}$201]Al$_2$O$_3$	
		(111)Si//(11$\bar{2}$4)Al$_2$O$_3$, [1$\bar{1}$0]Si//[$\bar{1}$100]Al$_2$O$_3$	
	MgO·Al$_2$O$_3$	(100)Si//(100) spinel,	
	(spinel)	(110)Si//(110) spinel,	
		(111)Si//(111) spinel	
SiC	Si	(110)SiC//(110)Si, [1$\bar{1}$1]SiC//[1$\bar{1}$1]Si	25)
		(111)SiC//(111)Si, [110]SiC//[110]Si	
Ge	CaF$_2$	(111)Ge//(111)CaF$_2$	26)
CdS	NaCl	(0001)CdS//(100)NaCl, [10$\bar{1}$0]CdS//[110]NaCl	27)
	Mica	(0001)CdS//(0001)mica, [10$\bar{1}$0]CdS//[10$\bar{1}$0]mica	
ZnO	α-Al$_2$O$_3$	(0001)ZnO/(0001)Al$_2$O$_3$, [11$\bar{2}$0]ZnO//[10$\bar{1}$0]Al$_2$O$_3$	28)
		(11$\bar{2}$0)ZnO/(01$\bar{1}$2)Al$_2$O$_3$, [0001]ZnO//[0$\bar{1}$11]Al$_2$O$_3$	

Table 2.6: Summaries of epitaxial relationship of single crystal substrates for various thin film materials.

Thin Film	Substrate	Epitaxial Relation	Ref.
AlN	α-Al$_2$O$_3$	(0001)AlN$/\!/(0001)$Al$_2$O$_3$, $[\bar{1}210]$AlN$/\!/[\bar{1}100]$Al$_2$O$_3$	29)
		$(11\bar{2}0)$AlN$/\!/(01\bar{1}2)$Al$_2$O$_3$, $[0001]$AlN$/\!/[01\bar{1}1]$Al$_2$O$_3$	30)
	Si	(0001)AlN$/\!/(111)$Si, $[11\bar{2}0]$AlN$/\!/[1\bar{1}0]$Si	31)
		(0001)AlN$/\!/(110)$Si, $[11\bar{2}0]$AlN$/\!/[\bar{1}10]$Si	
		(0001)AlN$/\!/(100)$Si, $[11\bar{2}0]$AlN$/\!/[011]$Si	
GaN	α-Al$_2$O$_3$	(0001)GaN$/\!/(0001)$Al$_2$O$_3$, $[\bar{1}210]$GaN$/\!/[\bar{1}100]$Al$_2$O$_3$	29)
		$(11\bar{2}0)$GaN$/\!/(01\bar{1}2)$Al$_2$O$_3$, (0001)GaN$/\!/[01\bar{1}1]$Al$_2$O$_3$	32)
GaAs	α-Al$_2$O$_3$	(111)GaAs$/\!/(0001)$Al$_2$O$_3$, $[1\bar{1}0]$GaAs$/\!/[11\bar{2}0]$Al$_2$O$_3$	29)
			33)
	MgO·Al$_2$O$_3$	(111)GaAs$/\!/(111)$spinel, $[01\bar{1}]$GaAs$[01\bar{1}]$spinel	29)
	(spinel)	(100)GaAs$/\!/(110)$spinel, $[011]$GaAs$/\!/[\bar{1}10]$spinel	34)
		(111)GaAs$/\!/(100)$spinel, $[01\bar{1}]$GaAs$/\!/[\bar{1}10]$spinel	
	BeO	(100)GaAs$/\!/(10\bar{1}1)$BeO	29)
		(111)GaAs$/\!/(0001)$BeO	35)
		(111)GaAs$/\!/(10\bar{1}1)$BeO	
GaP	α-Al$_2$O$_3$	(111)GaP$/\!/(0001)$Al$_2$O$_3$, $[1\bar{1}0]$GaP$/\!/[11\bar{2}0]$Al$_2$O$_3$	29)
	MgO·Al$_2$O$_3$	(111)GaP$/\!/(111)$ spinel, $[01\bar{1}]$GaP$/\!/[01\bar{1}]$ spinel	36)
	(spinel)		
	Si/α-Al$_2$O$_3$	(100)GaP$/(100)$Si$/(01\bar{1}2)$Al$_2$O$_3$	
InSb	NaCl	(0001)InSb$/\!/(001)$NaCl, $[10\bar{1}0]$InSb$/\!/[110]$NaCl	37)

Table 2.6: (continued) Summaries of epitaxial relationships of single crystal substrates for various thin film materials.

Thin Film	Substrate	Epitaxial Relation	Ref.
Pb-TiO$_3$ (PT)	α-Al$_2$O$_3$	(111)PT// (0001)Al$_2$O$_3$	
PLZT	MgO	(100)PLZT// (100)MgO	38)
	α-Al$_2$O$_3$	(111)PLZT// (0001)Al$_2$O$_3$	39)
	SrTiO$_3$	(100)PLZT// (100)SrTiO$_3$	
Li-NbO$_3$ (LN)	α-Al$_2$O$_3$	(0001)LN// (0001)Al$_2$O$_3$	40)
	LiTaO$_3$ (LT)	(0001)LN// (0001)LT	41)
Bi$_{12}$-TiO$_{20}$ (BTO)	Bi$_{12}$-GeO$_{20}$ (BGO)	(111)BTO// (111)BGO	42)

Table 2.6: (continued) Summaries of epitaxial relationships of single crystal substrates for various thin film materials.

Film properties	Evaluation methods	Remarks
Thickness	Optical color comparison	Transparent films on substrate, Simple: range 500 ∿ 15000A, accuracy 100 ∿ 200A.
	interferometer	Step and reflective coating: range 10 ∿ 20000A, accuracy 2 ∿ 30A.
	ellipsometry	Transparent films on substrate. range few A ∿ micron, accuracy ∿ A.
	Mechanical Stylus	Step required, Simple: range 20A to no limit, accuracy ∿ 10A.
Surface roughness	Mechanical stylus and/or optical microscope.	Simple.
	Scanning electron microscope.	Conductive coating needed for dielectric films: resolution ∿ 10A.
	Scanning tunneling microscope.	Conductive films: high resolution ∿ 1A. Atomic scale.

Table 2.7: Summaries of thin film evaluation methods.

Film properties	Evaluation methods	Remarks
Chemical composition	Inductively coupled plasma optical emission spectroscopy (ICP), Rutherford backscattering spectroscopy (RBS), Auger electron spectroscopy (AES), Electron probe micro-analysis (EPMA), X-ray photo-electron, spectroscopy (XPS) and/or Secondary ion mass spectroscopy (SIMS).	High sensitivity obtained by ICP and SIMS; detection limit $\sim$ 0.1 ppm. Non-destractive, quantitative analysis; by RBS and/or EPMA. Depth profile; by RBS, SIMS, and AES. Simple analysis; by ICP and/or EPMA.
Structure	Electron and/or X-ray diffraction analysis. X-ray photoelectron spectroscopy (XPS) and electron energy loss spectroscopy (EELS). Optical absorption.	Microstructural analysis, by TEM (transmission electron microscope). Amorphous films, by IR (infrared) absorption and/or EXAFS (extended X-ray absorption fine structure). Electronic states and valence states, by EELS and XPS, respectively. Refractive index, by ellipsometry.
Adhesion	Peeling method, scraching method, and pulling method.	For weak adhesion, film pelled off using a backing of adhesive tape. For strong adhesion (> 1 kg/cm^2), scraching and/or pulling methods.

Table 2.7: (continued) Summaries of thin film evaluation methods.

Film properties	Evaluation methods	Remarks
Stress	Disk method, bending-beam method, and X-ray diffraction method.	In the disk method, the film stress is measured by observing the deflection of the center of a circular plate; in the bending-beam method, by observing the deflection of a beam. Bending-beam method: Relation for stress σ, $\sigma = Ed\delta/3L^2(1-\nu)t$ E, ν, d, L denote Young modulus, Poisson's ratio, thickness, and length of substrate beam, respectively; film thickness.
Hardness	Micro-Vickers hardness measurement.	Measured at light load of identor. Hardness of substrate affects the film hardness for thin films. Extrapolated values at zero identor load give the true hardness for thin film.

Table 2.7: (continued) Summaries of thin film evaluation methods.

Film properties	Evaluation methods	Remarks
Wear & Friction	Wear test between film coated ball and iron plate. Sand blast method.	Strong adhesion onto the substrate is necessary for the wear test.
Electrical resistivity	Standard four terminals resistive measurements.	 Four terminals measurements. $= V/I \cdot \omega t/\ell$
Dielectric constant	Dielectric measurements at sandwitch structure; evaporated electrode/ dielectric film/ evaporated electrode on substrate, or interdigital electrodes (IDE).	Sandwitch structure: $\varepsilon^* = Ct/\varepsilon_0 S$, C, capacitance; t, film thickness; S, electrode area Interdigital electrodes: $\varepsilon^* \cong C/\varepsilon_0 n\ell K$, $K \equiv 6.5(D/L)^2$ $+ 1.08D/L + 2.37$ n, ℓ, D, L; numbers of pair, length, width, pitch of the IDE, respectively.

Table 2.7: (continued) Summaries of thin film evaluation methods.

Film properties	Evaluation methods	Remarks
Piezo-electricity	Admmittance measurements at sandwitch structure; evaporated electrode/ piezoelectric film/ evaporated electrode on fuzed quartz substrate.	Electromechanical coupling k_t $$k_t^2 \cong G_A X_C \pi Z_M / 4 Z_T,$$ G_A, X_C, conductance and capacitive reactance at antiresonant frequency, Z_M, Z_T; acoustic impedance of substrate and piezoelectric films, respectively.

Table 2.7: (continued) Summaries of thin film evaluation methods.

Techniques	Incident beam (particle)	Emitted beam (particle)	Spatial resolution (µm) dia.	Spatial resolution (µm) thick.	Detection limits (at. ppm)	Accuracy (%)	Elements	Other features
X-ray fluoresence spectroscopy	X-rays	X-rays	10000	30	1~100	±1	Z≥9	Quantitative Non-destructive
Electron microprobe analysis	Electrons	X-rays	1	1	0.01~0.1%	±2	Z≥4	Quantitative Non-destructive
Particle induced x-ray emission	Ions	X-rays	1	1	1	--	Z≥4	Elemental analysis
Rutherford back-scattering spectroscopy	Ions	Ions	3	0.03	0.1~0.01%	±3	Z≥5	Quantitative Non-destructive Depth profile
Ion scattering spectroscopy	Ions	Ions	1000	0.0003	0.1~1%	±20	Z≥6	Semi-quantitative Depth profile
Secondary ion mass spectrometry	Ions	Ions	1	0.003	0.1~100	±20		Semi-quantitative Depth profile Conductor
Auger electron spectroscopy	Electrons	Electrons	0.05	0.003	0.1~1%	±20	Z≥3	Semi-quantitative Depth profile Conductor
Electron energy loss spectroscopy	Electrons	Electrons	0.01	0.05		--		Elemental analysis Elecronic states
X-ray photoelectron spectroscopy	X-rays	Electrons	150	0.003	0.1~1%	±20	Z≥3	Semi-quantitative Valence states Non-destructive
Reflection high energy electron diffraction	Electrons	Electrons	100Å	100Å				Surface structure
Low energy electron diffraction	Electrons	Electrons	300Å	3Å				Surface structure
Transmission electron microscopy	Electrons	Electrons	100Å	2Å				Surface/interface structure
Scanning electron microscopy	Electrons	Electrons	20Å	30Å				Surface topograph Conductor
Scanning tunneling microscopy	Distance (1~2A)	Tunneling current (1-10nA)	3Å	0.01Å				Surface topograph Conductor Bias (1mV~1V)

Table 2.8: Summary of surface and thin film analysis methods.

2.4 REFERENCES

1. Bunshah, R.F., (ed.) Deposition Technologies for Films and Coatings, New Jersey, Noyes Publications (1982). Chopra K.L., and Kaur, I., Thin Film Device Applications, New York, Plenum Press (1983).

2. Kaur, I., Pandya, D.K., and Chopra, K.L., J. Electrochem. Soc. 127: 943 (1980).

3. Chopra, K.L., Randlett, M.R., and Duff, R.H., Philos, Mag. 16: 261 (1967).

4. L.Esaki, Proc. 6th Int. Vacuum Congr., Kyoto, Jpn. J.Appl.Phys., 13: Suppl. 2, Pt. 1, 821-828 (1974)..

5. Tokumitsu, E., Kudou, Y., Konagai, M., and Takahashi, K., J. Appl. Phys., 55: 3163 (1984).

6. Wehner G.K., and Anderson, G.S., Handbook of Thin Film Technology, 3-1, (L. Maissel and R. Glang, eds.) New York, McGraw Hill (1970).

7. F.M.Penning, U.S. Patent 2,146,025 (Feb. 1935).

8. Wasa, K., and S.Hayakawa, S., Rev. Sci. Instrum. 40: 693 (1969).

9. Chapin, J.S., Res./Dev, 25: 37 (1974).

10. Chopra K.L. and Randlett, M.R., Rev. Sci. Instr., 38: 1147 (1967).

11. Vossen, J.L. and Kern, W. (eds.) Thin Film Processes, New York, Academic Press (1978).

12. Kitabatake, M., Wasa, K., J. Appl. Phys., 58: 1693 (1987).

13. Behrisch, R. (ed.) Sputtering by Particle Bombardment I, II Berlin, Springer-Verlag (1981), (1983).

14. Mattox, D.M., J. Vac. Sci. Technol., 10: 47 (1973).

15. Bunshah R.F., and Raghuram, A.C., J. Vac. Sci. Technol., 9: 1385 (1972).

16. Takagi, T., Yamada, I., and Sasaki, A., J. Vac. Sci. Technol., 12: 1128 (1975), Takagi, T., in Ionized Cluster Beam Deposition and Epitaxy, New Jersey, Noyes Publications (1988).

17. Nishizama, J., Kurabayashi, T., Abe, H., and Sakurai, N., J. Vac. Sci. Technol., A5: 1572 (1987).

18. I.H.Khan, in Handbook of Thin Film Technology, (L. Maissel and R. Glang, eds.) 10-1, McGraw Hill, New York (1970).

19. Spiegel, K., Surface Sci., 7: 125 (1967).

20. Stirland, D.J., Appl. Phys. Lett., 8: 326 (1966).

21. Schlier, R.E., Fransworth, H.E., J. Phys. Chem. Solids, 6: 271 (1958).

22. Gonzalez, C., and Grunbam, E., Proc. 5th Intern. Conf. Electron Microscopy, Vol.1, p.DD-1, Academic Press Inc., New York (1962).

23. Francombe, H.M., in The Use of Thin Films in Physical Investigations p65, (G.E. Anderson, ed.), Academic Press, New York (1966).

24. Maissel, L.I., and   Glang R., (eds.)   Handbook of Thin Film Technology 10-12, McGraw Hill, New York (1970).

25. Khan, I.H., Summergrad, R.N., Appl.Phys. Lett., II: 12 (1967).

26. Pundsack, A.L., J. Appl. Phys., 34: 2306 (1963).

27. Chopra, K.L. and Khan, I.H., Surface Sci., 6: 33 (1967).

28. Mitsuyu, T., Ono, S., and Wasa, K., J. Appl. Phys., 51: 2464 (1980).

29. Wang, C.C., McFarlane III, S.H., Thin Solid Films 31: 3 (1976).

30. Morita, M., Uesugi, N., Isogai, S., Tsubouchi, K., Mikoshiba, N., Jpn. J. Appl. Phys., 20: 17 (1981).

31. Morita, M., Isogai, S., Shimizu, N., Tsubouchi, K., Mikoshiba, N., Jpn. J. Appl. Phys., 20: L173 (1981).

32. Thorsen, A.C., Manasevit, H.M., J. Appl. Phys., 42: 2519 (1971).

33. Manasevit, H.M., Simpson, W.I. J. Electrochem. Soc., 116: 1725 (1969).

34. Wang, C.C., Dougherty, F.C., Zanzucchi, P.J., McFarlane III, S.H., J. Electrochem. Soc., 121: 571 (1974).

35. Thorsen, A.C., Manasevit, H.M., Harada, R.H., Solid-State Electron., 17: 855 (1974).

36. Wang, C.C., McFarlane III, S.H., J. Cryst. Growth, 13-14: 262 (1972).

37. Khan, I.H., Surface Sci., 9: 306 (1968).

38. Okuyama, K., Usui, T., Hamakawa, Y., Appl. Phys.,21: 339 (1980).

39. Ishida, M., Tsuji, S., Kimura, K., Matsunami, H., Tanaka, T., J. Crystal Growth, 45: 393 (1978).

40. Takada, S., Ohnishi, M., Hayakawa, H., Mikoshiba, N., Appl. Phys. Lett., 24: 490 (1974).

41. Miyazawa, S., Fushimi, S., Kondo, S, Appl. Phys.Lett., 26: 8 (1978).

42. Mitsuyu, T., Wasa, K., Hayakawa, S., J. Crystal Growth, 41: 151 (1977).

43. Wasa, K., Nagai, T., Hayakawa, S., Thin Solid Films, 31: 235 (1976).

44. Nagai, T., Wasa, K., Hayakma, S. J. Materials Sci., 11: 1509 (1976).

45. Farnell, G.W., Cermak, I.A., Silvester, P., Wong, S.K., IEEE Trans. Sonics Ultrason., SU-17: 188 (1970).

46. Bahr, A.J., Court, I.N., J.Appl. Phys., 39: 2863 (1968).

47. Huttemann, R.D., Morabito, J.M., Stieidel, C.A., Gerstenberg, D., Proc. 6th Int. Vacuum Congr., Kyoto,(1974), Jpn. J. Appl. Phys., 13: Suppl. 2, Pt-1, p.513 (1974).

3

SPUTTERING PHENOMENA

Sputtering was first observed in a dc gas discharge tube by Grove in 1852. He discovered the cathode surface of the discharge tube was sputtered by energetic ions in the gas discharge, and cathode materials were deposited on the inner wall of the discharge tube.

At that time sputtering was regarded as an undesired phenomena since the cathode and grid in the gas discharge tube were destroyed. Today, however, sputtering is widely used for surface cleaning and etching, thin film deposition, surface and surface layer analysis, and sputter ion sources.

In this chapter, the fundamental concepts of the various sputtering technologies are described. The energetic particles in sputtering may be ions, neutral atoms, neutrons, electrons or photons. Since most relevant sputtering applications are performed under bombardment with ions, this text deals with that particular process.

3.1 SPUTTER YIELD

The sputter yield S, which is the removal rate of surface atoms due to ion bombardment, is defined as the mean number of atoms removed from the surface of a solid per incident ion and is given by

$$S = \frac{\text{atoms removed}}{\text{incident ions}} \,. \tag{3.1}$$

Sputtering is caused by the interactions of incident particles with target surface atoms. The sputter yield will be influenced by the following factors;

1. energy of incident particles

2. target materials

3. incident angles of particles

4. crystal structure of the target surface

The sputter yield S can be measured by the following methods

1.   weight loss of target

2.   decrease of target thickness

3.   collection of the sputtered materials

4.   detection of sputtered particles in flight

The sputter yield is commonly measured by weight loss experiments using a quartz crystal oscillator microbalance (QCOM) technique. Surface analysis techniques including Rutherford back scattering spectroscopy (RBS) are available for measuring the change in thickness or composition of targets on an atomic scale during sputtering. RBS is essentially non-destructive and the dynamic sputter yield is determined with a priori accuracy of some 10%. Scanning electron microscope (SEM) and stylus techniques are used for the measurement of minute change in target thickness. These techniques need an ion erosion depth in excess of around 0.1μ. The QCOM technique is sensitive probing method with sub-monolayer resolution (1).

Both electron and proton probe beam techniques are also used successfully in-situ dynamic and absolute yield determinations. Auger electron spectroscopy (AES) could also be used for the determination of monolayer thickness. Proton-induced X-ray emission (PIXE) with proton energy of 100 to 200keV (2) and electron-induced X-ray emission with electron energy of around 10keV are also used for the sputter yield measurement (3). The PIXE technique could quantify both initial surface impurities as well as the pure sputter yield of the target.

3.1.1 Ion Energy

Figure 3.1 shows a typical variation of the sputtering yield with incident ion energy. The figure suggests:

1.   In a low energy region, a threshold energy exists for sputtering.

2.   The sputter yield shows maximum value in a high energy region.

Hull first observed the existence of the sputtering threshold in 1923. He found that the Th-W thermionic cathode in gas rectifier tubes was damaged by bombardment with ions when the bombarding ion energy exceeded a critical value which was in the order of 20 to 30 eV (4). The sputtering threshold has been studied by many workers because it is probably related to the mechanism of sputtering. Threshold values obtained by these workers ranged from 50 to 300 eV (5,6). Their results were somewhat doubtful because the threshold energy was mainly determined by measurements of small weight loss from the cathode in the range of 10^{-4} atoms/ion. The threshold energy is very sensitive to contamination of the cathode surface. In addition, the incident angle of ions and the crystal orientation of cathode materials also change the threshold values.

In 1962, Stuart and Wehner skillfully measured reliable threshold values by the spectroscopic method (7). They observed that threshold values are in the order 15 to 30 eV, similar to the observations of Hull, and are roughly four times the heat of sublimation of cathode materials.

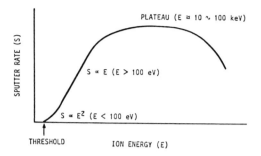

Figure 3.1: Variations of sputter yield with incident ion energy.

A gas discharge tube used by Stuart and Wehner is shown in Fig. 3.2 (7). The target is immersed in plasma generated by a low-pressure mercury discharge. In a noble gas discharge the mercury background pressure is about 10^{-5} Torr or less. The temperature of the target is kept at about $300°C$ so as to reduce condensation of the mercury vapor. The sputtered target atoms are excited in the plasma and emit the specific spectrum. The sputter yields at low energy are determined by the intensity of the spectral line. This technique eliminates the need for very sensitive weight measurements.

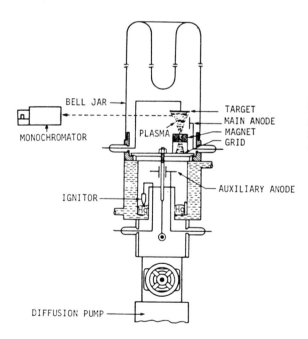

Figure 3.2: Experimental apparatus for the measurements of the threshold energy (Stuart, Wehner, 1962 (7)).

Typical sputter yields in a low energy region measured by Stuart and Wehner are shown in Fig. 3.3. The threshold values determined in the sputter yields are on the order of 10^{-4} to 10^{-5} atoms/ion. Table 3.1 summarized the sputtering threshold energy measured by the spectroscopic method for various target materials (7,8).

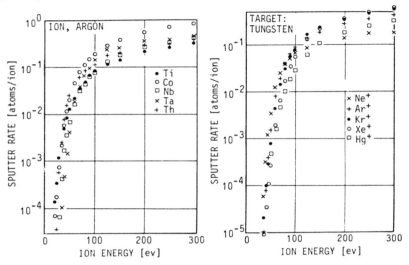

Figure 3.3: Sputter yield in a low energy region (Stuart, Wehner, 1962 (7)).

(eV)

	Ne	Ar	Kr	Xe	Hg	H
Be	12	15	15	15	—	—
Al	13	13	15	18	18	—
Ti	22	20	17	18	25	4.40
V	21	23	25	28	25	5.28
Cr	22	22	18	20	23	4.03
Fe	22	20	25	23	25	4.12
Co	20	25	22	22	—	4.40
Ni	23	21	25	20	—	4.41
Cu	17	17	16	15	20	3.53
Ge	23	25	22	18	25	4.07
Zr	23	22	18	25	30	6.14
Nb	27	25	26	32	—	7.71
Mo	24	24	28	27	32	6.15
Rh	25	24	25	25	—	5.98
Pd	20	20	20	15	20	4.08
Ag	12	15	15	17	—	3.35
Ta	25	26	30	30	30	8.02
W	35	33	30	30	30	8.80
Re	35	35	25	30	35	—
Pt	27	25	22	22	25	5.60
Au	20	20	20	18	—	3.90
Th	20	24	25	25	—	7.07
U	20	23	25	22	27	9.57
Ir		(8)				5.22

Table 3.1: Sputtering threshold data (Stuart, Wehner, 1962 (7), Harrison, Magnuson, 1961 (8)).

H, Heat of sublimation

The table suggests that there is not much difference in threshold values. The lowest value, which is nearly equal to four times the heat of sublimation, is observed for the best mass fit between target atom and incident ion. The higher threshold energy is observed for poor mass fits.

The threshold energy also strongly depends on the particular sputtering collision sequence involved. High threshold energy (i.e. $E_{th}/U_o > 10$ for $Ar^+ \to Cu$ where E_{th} denotes the threshold energy; U_o, the heat of sublimation) will be expected in the collision sequence where the a primary recoil produced in the first collision is ejected directly. Lower values will be observed for the multiple sputtering collisions. An incident angle of around 40 to 60° offers the minimum threshold energy (i.e. $E_{th}/U_o \simeq 2$ for $Ar^+ \to Cu$) under the multiple sputtering collisions (9).

The sputter yield varies with the incident ion energy E. In the low energy region near the threshold the S obeys the relation $S \propto E^2$ as seen in Fig. 3.3. At the energy region in the order of 100 eV, $S \propto E$ (6,10). In this energy region, the incident ions collide with the surface atoms of the target, and the number of displaced atoms due to the collision will be proportional to the incident energy.

At higher ion energies of 10 to 100 keV, the incident ions travel beneath the surface and the sputter yields are not governed by the surface scattering but by the scattering inside of the target. Above 10 keV the sputter yields will decrease due to energy dissipation of the incident ions deep in the target.

Maximum sputter yields are seen in the ion energy region of about 10 keV. Figure 3.4 summarizes the energy dependence of the yield, i.e., the sputter rate, as reported by Sigmund (11).

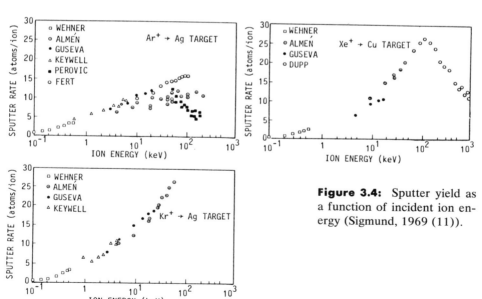

Figure 3.4: Sputter yield as a function of incident ion energy (Sigmund, 1969 (11)).

3.1.2 Incident Ions, Target Materials

Sputter yield data have been extensively accumulated in relation to gas discharges, sputter deposition, etching, surface analysis, and radiation damage. At first, sputter yields were measured in the cold cathode discharge tube (12,13). But these measurements did not offer reliable data because the incident ions and sputtered atoms frequently collide with discharge gas molecules in the cold cathode discharge tube.

Laegreid and Wehner, in 1959, accumulated the first reliable data of sputter yields in a low gas pressure discharge tube (14). At present, this yield data is still widely used for sputtering applications. Figure 3.5 shows Wehner's sputtering stand (6,15). The system is based on a mercury discharge tube. The discharge is maintained at low gas pressure by a thermionic mercury cathode. The discharge gas is 1×10^{-3} Torr for Hg, and $2 \simeq 5 \times 10^{-3}$ Torr for Ar, 4×10^{-2} Torr for Ne. The sputter yield S is determined by

$$S = 10^5(W/AIt), \tag{3.2}$$

where W denotes the weight loss of target during the sputtering time t with the ion current I to the target, and A denotes the atomic number of the target materials.

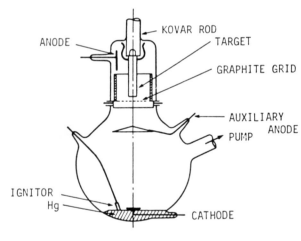

Figure 3.5: Wehner's experimental apparatus for the measurement of sputter yield (Wehner, 1957 (6)).

Typical results are shown in Fig. 3.6 and Table 3.2. Polycrystalline targets were used in the measurements so the effects of the crystal orientation could be neglected. It is noted that this yield data correspond to $S/(1 + \gamma)$, where γ is the secondary electron emission coefficient of the target materials. The values of γ for various combinations of incident ions and target materials are shown in Table 3.3 (16). For sputter deposition, the range of the incident ion energy is below 1,000 eV where the $\gamma \simeq 0.1$ for Ar+. The error of Wehner's data will be typically less than 10% even if the effects of the secondary electron emission are not taken into consideration.

Recent PIXE techniques suggest that the sputter yield of Cr, S(Cr), measured by Wehner is larger than S(Cr) measured by the PIXE. Sartwell has measured that S(Cr)=0.93 atoms/ion at 1.0 keV Ar+ bombardment measured by PIXE (2). Wehner has reported a value of S(Cr) = 1.18 atoms/ion at 500eV. The Wehner's S(Cr) indicated in Fig. 3.6 and Table 3.2 are 1.5 times larger that the PIXE measurements (2).

As indicated in Fig. 3.6, sputter yields vary periodically with the element's atomic number. Comparing various materials, the yields increase consistently as the d shells are filled, with Cu, Ag, and Au having the highest yields. Periodicity is also observed in sputtering thresholds (17). Sigmund offered theoretical consideration in detail (11).

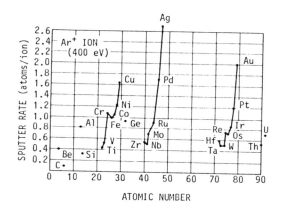

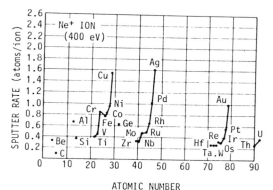

Figure 3.6: Sputter yield vs. atomic number for the impingment of Ar+, Ne+, and Hg+ (Laegreid, Wehner, 1961 (11)).

Target	Ne+				Ar+			
	100 (eV)	200 (eV)	300 (eV)	600 (eV)	100 (eV)	200 (eV)	300 (eV)	600 (eV)
Be	0.012	0.10	0.26	0.56	0.074	0.18	0.29	0.80
Al	0.031	0.24	0.43	0.83	0.11	0.35	0.65	1.24
Si	0.034	0.13	0.25	0.54	0.07	0.18	0.31	0.53
Ti	0.08	0.22	0.30	0.45	0.081	0.22	0.33	0.58
V	0.06	0.17	0.36	0.55	0.11	0.31	0.41	0.70
Cr	0.18	0.49	0.73	1.05	0.30	0.67	0.87	1.30
Fe	0.18	0.38	0.62	0.97	0.20	0.53	0.76	1.26
Co	0.084	0.41	0.64	0.99	0.15	0.57	0.81	1.36
Ni	0.22	0.46	0.65	1.34	0.28	0.66	0.95	1.52
Cu	0.26	0.84	1.20	2.00	0.48	1.10	1.59	2.30
Ge	0.12	0.32	0.48	0.82	0.22	0.50	0.74	1.22
Zr	0.054	0.17	0.27	0.42	0.12	0.28	0.41	0.75
Nb	0.051	0.16	0.23	0.42	0.068	0.25	0.40	0.65
Mo	0.10	0.24	0.34	0.54	0.13	0.40	0.58	0.93
Ru	0.078	0.26	0.38	0.67	0.14	0.41	0.68	1.30
Rh	0.081	0.36	0.52	0.77	0.19	0.55	0.86	1.46
Pd	0.14	0.59	0.82	1.32	0.42	1.00	1.41	2.39
Ag	0.27	1.00	1.30	1.98	0.63	1.58	2.20	3.40
Hf	0.057	0.15	0.22	0.39	0.16	0.35	0.48	0.83
Ta	0.056	0.13	0.18	0.30	0.10	0.28	0.41	0.62
W	0.038	0.13	0.18	0.32	0.068	0.29	0.40	0.62
Re	0.04	0.15	0.24	0.42	0.10	0.37	0.56	0.91
Os	0.032	0.16	0.24	0.41	0.057	0.36	0.56	0.95
Ir	0.069	0.21	0.30	0.46	0.12	0.43	0.70	1.17
Pt	0.12	0.31	0.44	0.70	0.20	0.63	0.95	1.56
Au	0.20	0.56	0.84	1.18	0.32	1.07	1.65	2.43 (500)
Th	0.028	0.11	0.17	0.36	0.097	0.27	0.42	0.66
U	0.063	0.20	0.30	0.52	0.14	0.35	0.59	0.97

Table 3.2: Sputter yield (Laegried, Wehner, 1962 (15)).

Target Materials	Incident Ion	Ion Energy (eV)		
		200	600	1000
W	He+	0.524	0.24	0.258
	Ne+	0.258	0.25	0.25
	Ar+	0.1	0.104	0.108
	Kr+	0.05	0.054	0.058
	Xe+	0.016	0.016	0.016
Mo	He+	0.215	0.225	0.245
	He++	0.715	0.77	0.78
Ni	He+		0.6	0.84
	Ne+			0.53
	Ar+		0.09	0.156

Table 3.3: Secondary electron coefficients, γ (Brown, 1959 (16)).

3.1.3 Angle of Incidence Effects

Sputter yields vary with the angle of incident ions. Fetz studied the influence of incident ions in 1942 and, later, Wehner studied the topic in detail (18,19). Metals such as Au, Ag, Cu, and Pt which have high sputtering yields show a very slight "angle effect". Fe, Ta, and Mo having low sputtering yields show a very pronounced angle effect. The yield increases with the incident angle and shows a maximum at angles between 60° and 80°, while it decreases rapidly for larger angles. The influence of the angle is also governed by the surface structure of the target. Theoretical studies have been done by several workers (20). Recent results obtained by Bay and Bohdansky are shown in Fig. 3.7 (21).

Figure 3.8 shows the angular distribution of sputtered atoms for the oblique incidence of bombarding ions measured by Okutani et al (22). The atoms are ejected preferentially in a forward direction.

The angular distributions of sputtered atoms for the normal incidence ion bombardment were studied in various ranges of incident ion energy. Seeliger and Sommermeyer measured them in a high energy region of 10 keV. These experiments suggested that the angular distribution was governed by Knudsen's cosine law which was observed in a thermal evaporation process (23). Wehner and Rosenberg have measured the angular distribution in the lower energy region of 100 to 1,000 eV in a low pressure mercury discharge tube which is shown in Fig. 3.9. The target strip is mounted in the center of the glass cylinder which holds the collecting glass ribbon on its inside wall. The density of material sputtered over the entire 180° of the ribbon gives the angular distribution. Typical results are shown in Fig. 3.10 (19). It suggests that the angular distribution is "under cosine", i.e., much more material is ejected to the sides than in the direction normal to the target surface. The distribution will approach a cosine distribution at higher ion energies. Mo and Fe show a more pronounced tendency to eject to the sides than Ni or Pt. At energies higher than 10 keV the distribution shows "over cosine".

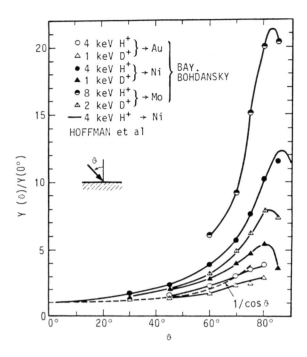

Figure 3.7: Sputter yield vs. incident angle of bombarding ions (Anderson, Bay, 1981 (21)).

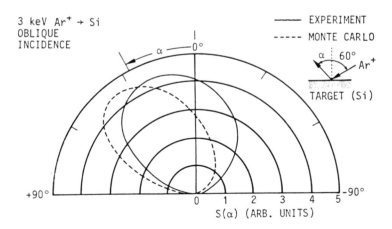

Figure 3.8: Angular distributions of sputtered Si atoms for 3keV Ar⁺ ion bombardment at incident angle of 60° (Okutani et al, 1980 (22)).

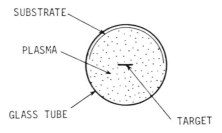

Figure 3.9: Experimental sputtering apparatus for the measurements of angular distributions (Wehner, Rosenberg, 1960 (19)).

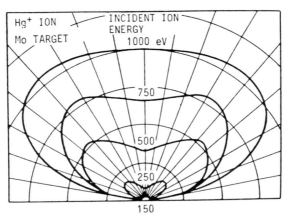

Figure 3.10: Angular distributions of sputtered particles from polycrystal target (Wehner, Rosenberg, 1960 19)).

Angular distributions are related to the sputtering mechanism and also considered in several applications including secondary ion mass spectrometry (SIMS), sputter deposition, and sputter etching. Recently angular distributions have been studied in detail using ion beam sputtering systems (24).

3.1.4 Crystal Structure of Target

It is well known that the sputtering yield and the angular distribution of the sputtered particles are affected by the crystal structure of the target surface. As described in the previous section, the angular distribution may be either under cosine law or over cosine law when the target is composed of polycrystalline materials. Non uniform angular distribution is often observed from the single crystal target.

Wehner studied the nonuniform angular distribution of single crystals in detail and found deposited patterns appeared (6). He suggests that near threshold the sputtered atoms are ejected in the direction of close-packed atoms. For instance, in fcc Ag the close-packed direction corresponds to <110>. When an fcc (111) Ag target is sputtered a three-fold symmetrical patterns appear, since there are three close-packed directions in the (111) plane.

Table 3.4 shows a summary of the sputtered pattern for single crystal targets.

At higher ion energy additional atoms are freed from the more numerous positions where neighbor atoms may interfere with the direction of close-packed rows. This causes deviations from these directions.

Target	Crystal Structure	Direction of Sputtered Pattern
Ag	fcc	<110>
W, Mo, α-Fe	bcc	<111>
Ge	diamond	<111>
Zn, Ti, Re	hexagonal	$< 11\bar{2}0 >$, $< 20\bar{2}3 >$

Table 3.4: Sputtered pattern for single crystal targets.

Measurements of Ar^+ incident onto fcc crystals such as Cu, Ag, Au, and Al confirm the general features; the sputter yields $S_{111} > S_{poly} > S_{110}$ for incident energies of a few keV (25). At low incident energies $< 100eV$ the ordering of the sputter yields becomes $S_{110} > S_{100} > S_{111}$ (6). The theoretical approaches suggest that the ordering of the sputter yields relates to that of the binding energy $U_{111} > U_{100} > U_{100}$ (26).

The sputter yields of hcp crystals for a few keV Ar^+ bombardment exhibit: $S_{0001} > S_{1010} > S_{1120}$ for Mg, $S_{1010} > S_{0001} > S_{1120}$ for Zr, and $S_{0001} > S_{1010} > S_{1120}$ for Zn and Cd. The ordering of these sputtering yields relates the ordering of the interatomic distance t for the [0001] and [1010] direction with the value c/a: $t_{0001} > t_{1010} > t_{1120}$ for Cd and Zn, $t_{1010} > t_{0001}$ for Mg and Zr (25). Typical experimental results are shown in Fig. 3.11 (27).

It is also interesting that the angular dependence of the sputter yield for moncrystalline target shows distinct peaks for ejection directions. The ejection direction can be crystallographically characterized by low Miller indices as first pointed out by Wehner (6). Several models have been considered to explain the angular dependence of the sputter yield for the moncrystalline target including the transparency model by Fluit (28) and the channeling model by Onderdelinden (29). Fig. 3.12 shows the angular dependence of the sputter yield measured at 27kev Ar^+ on a (111) Cu crystal turned around the (112) axis with theoretical values due to the channeling model (30).

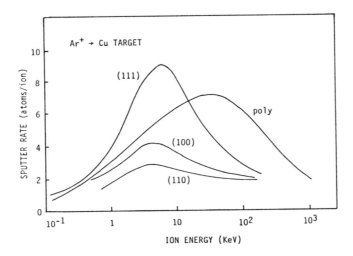

Figure 3.11: Energy dependance of the sputter yields of Ar⁺ on the (110), (100) and (111) planes of Cu (Roosendaal, 1983 (27)).

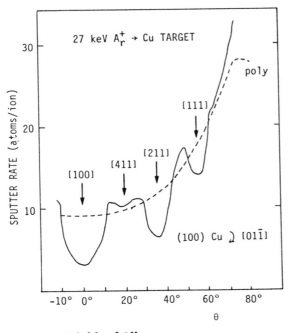

Figure 3.12: Angular dependance of the sputter yields for Ar⁺ on the (111) Cu rotated around the (112) axis (Elich, 1972) (30).

3.1.5 Sputter Yields of Alloys

Experience has shown that the chemical composition of the deposited film sputtered from an alloy target is very close to that of the target. This suggests that sputtering is not governed by thermal processes but by a momentum transfer process. When the temperature of the target is so high that the composition of the alloy target changes due to the thermal diffusion, the resultant films show a different chemical composition. Under even higher substrate temperature the resultant films also show a different chemical composition because of the reevaporation of the deposited films.

Several alloys are routinely sputtered including Permalloy (81Ni-19Fe) and Ni-Cr alloys. Flur and Riseman found that films sputtered from a Permalloy target had the same composition as the target (31).

Patterson and Shirn suggested that the composition of the Ni-Cr target was preserved in the sputtered films (32). Sputtering conditions will alter the composition of the sputtered films. The detailed composition of the sputtered films measured was different between several workers due to measuring instruments.

The nature of changes in the alloy target is significant. Patterson showed that variation of the surface composition during sputtering deposition for a binary alloy target A, B was expressed by the following relations (32);

$$A = [A_0 - \frac{S_B N_0 A_0}{A_0 S_B + B_0 S_A}] \exp - (\frac{A_{0SB} + B_0 S_A}{N_0^2} Ft) + \frac{S_B N_0 A_0}{A_0 S_B + B_0 S_A}, \qquad (3.3)$$

$$B = [B_0 - \frac{S_A N_0 B_0}{A_0 S_B + B_0 S_A}] \exp - (\frac{A_{0SB} + B_0 S_A}{N_0^2} Ft) + \frac{S_A N_0 B_0}{A_0 S_B + B_0 S_A}, \qquad (3.4)$$

where A_0 and B_0 denote the surface density of the composition A, B at the initial stage of the sputtering, respectively. S_A and S_B are the sputter yields of A, B, respectively. N_0 is the atomic density of the alloy target surface, F is the ion current density at the target surface, and t is the sputtering time.

These relations are obtained under the assumption that values S_A and S_B for the alloy are the same to those of each element A and B, respectively. We know these relations are useful when each element A and B show small differences in their atomic weights. A typical example is Ni-Cr alloys.

Eqns. (3.3) and (3.4) suggest that the surface composition of the target will change in the sputtering time. The time constant for this change is

$$(\frac{A_0 S_B + B_0 S_A}{N_0^2} F)^{-1}$$

As t → ∞,

$$A \rightarrow \frac{A_0 S_B N_0}{A_0 S_B + B_0 S_A}, \qquad (3.5)$$

$$B \rightarrow \frac{B_0 S_A N_0}{A_0 S_B + B_0 S_A},$$

(3.6)

and A and B approach a constant value.

Let us consider, for example, the sputtering of Ni-20Cr $(A_0 = 0.8N_0, B_0 = 0.2N_0)$ by 600 eV Ar ions, in which case the sputter yield for Ni(S_A) is 1.5 atoms per ion, and Cr$(S_B) = 1.3$. The surface density, N_0 is about 2×10^{15} atcm^{-2}. If F is 6×10^{15} cm^{-2} sec^{-1} (1mA/cm^2), the time constant

$$[\frac{A_0 S_B + B_0 S_A}{N_0{}^2} F]^{-1}$$

becomes 250 msec. The calculated time constant is fairly shorter than the sputtering time for the deposition. It can be shown from Eqns. (3.5) and (3.6) that the surface of the target will change from 80Ni-20Cr to 78Ni-22Cr during one time constant.

The instantaneous film compositions for A and B are expressed by $\int_0^t S_A A dt$ and $\int_0^t S_B B dt$, respectively. Then the ratio of the each composition of the sputtered film at $t = \infty$, (A_S/B_S) becomes

$$A_S/B_S = A_0/B_0.$$

(3.7)

Eqn. (3.7) suggests that the sputtered film should have the same composition to the target in the steady state when the solid state diffusion in the target is neglected. At ordinary temperatures (a few hundred degrees centigrade), such diffusion will be unimportant (32).

Recently Liau has found the surface change of the target composition for Pt and Si alloy is not governed by Eqns (3.5) and (3.6) (33). He studied the surface composition by Rutherford backscattering techniques and showed that the Si, which has a small elemental sputter yield has been preferentially sputtered off. After sputtering, the Pt, which has a high elemental yield, was found to be enriched in the surface of the target.

Table 3.5 shows the steady state composition of the surface layers for 40 keV Ar ion sputtering for various binary alloys allied to Pt/Si. It was suggested that these phenomena were chiefly observed in the binary alloys which were composed of light components and heavy components. The heavy components were generally enriched in the surface layer.

Haff has studied the phenomena of surface layer enrichment and suggested that is was related to the collision cascade between two species of binary alloys in the surface layer (34). Tarng and Wehner showed that the surface change of Cu-Ni alloy is governed by Eqns. (3.5) and (3.6), although Cu is much heavier than Ni (35).

As described above the compositional change of the target surface is complicated. The composition of the sputtered alloy films, however, is generally equal to the target composition when the target is cooled during deposition (36). Table 3.6 shows some examples of the composition of alloy films sputtered from conventional magnetron sputtering (37).

Bulk	Surface
$Au_{0.19}A$	$Au_{0.23}A$
Cu_3Al	$Cu_3Al_{1.1}$
Au_2Al	$Au_{3.3}Al$
$AuAl_2$	$Au_{1.3}Al_2$
Pt_2Si	$Pt_{3.3}Si$
PtSi	$Pt_{2.1}Si$
NiSi	$Ni_{1.6}Si$
InP	same as bulk
GaP	same as bulk
GeSi	same as bulk
Ta_2O_5	$Ta_{4.5}O_5$

Table 3.5: Variations of surface compositions under ion bombardment with $40keVAr^+$ ions (Liau, Brown, (1977)(33)).

Materials	Target Composition	Film Composition
Cu	3.9 - 5%	3.81%
Si	0.5 - 1%	0.86%
Mn	0.4 - 1.2%	0.67%
2014 Al alloy Mg	0.2 - 0.8%	0.24%
Fe	1.0%	0.21%
Cr	0.10%	0.02%
Zn	0.25%	0.24%
(Al + Cu + Si) Cu	4%	3.4%
Si	2%	2.8%
(Al + Si) Si	2%	$2\% \pm 0.1\%$

Table 3.6: Compositions of Al alloy thin films deposited by magnetron sputtering (Wilson, Terry, 1976 (36)).

3.2 SPUTTERED ATOMS

3.2.1 Features of Sputtered Atoms

In a conventional sputtering system sputtered atoms are generally composed of a neutral, single atoms of the target material when the target is sputtered by bombardment with ions having a few hundred electron volts. These sputtered atoms are partially ionized, i.e., a few percent of the sputtered atoms, in the discharge region of the sputtering system.

Woodyard and Cooper have studied the features of sputtered Cu atoms under bombardment with 100 eV Ar ions, using a mass spectrometer. They found that 95% of the sputtered atoms are composed of single Cu atoms and the remaining 5%, Cu_2 molecules (38).

Under higher incident ion energy, clusters of atoms are included in the sputtered atoms. Herzog showed the sputtered atoms were comprised of clusters of Al_7, when the Al target was sputtered by 12 keV Ar ions. Under bombardment with Xe ions, clusters of Al_{13} were detected.

For an alloy target, the features of the sputtered atoms are similar to those of the single element target. Under low incident ion energy most of the sputtered atoms are composed of the single element of the alloy. Clusters are predominant when the ion energy is higher than 10 keV.

3.2.2 Velocity and Mean Free Path

3.2.2.1 Velocity of Sputtered Atoms:
The average energy of sputtered neutral atoms is much higher than that of thermally evaporated atoms in a vacuum. This phenomena was first observed by Guenthershulze (39), Mayer (40), and Spron (41).

Wehner has studied the velocity of the sputtered atoms in detail. He measured the velocity by a quartz balance shown in Fig. 3.13 (42). A low pressure Hg plasma is maintained between an anode and a Hg pool cathode. A quartz helix balance with a little quartz pan are suspended in the upper part of the tube. When the sputtered atoms deposit on the underside of the pan, the atoms exert a force $(dM/dt)\bar{v}_z$ which displaces the pan upward by a certain distance, where dM/dt denotes the mass per second arriving, $\bar{v}_z$ is the average velocity component normal to the pan surface. The continuous deposition increases the weight of the pan with time, and the pan will return to its original position. When one measures the time interval, t, required for the pan to return to its original position, the $\bar{v}_z$ is estimated from the relation $\bar{v}_z = tg$ $(dM/dt)\bar{v}_z = (dM/dt)tg$, $g = 981 cm/ sec^2$) without knowledge of dM/dt.

The quartz balance method is also used for determining the average velocity of vacuum evaporated particles. Wehner has estimated the average velocity by this method and suggested that the average velocity of Pt, Au, Ni, W sputtered by Hg ions of a few hundred eV is of the order 3 to $7x10^5$ cm/sec corresponding to the average kinetic energy of 10 to 30 eV. The kinetic energy is more than 100 times higher than thermal evaporation energy.

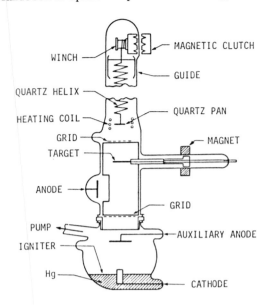

Figure 3.13: A quartz balance for the measurements of the velocity of the sputtered atoms (Wehner, 1959 (42)).

Both the calorimetric method and time-of-flight method are also used for determining the average energy of the sputtered atoms. In the calorimetric method, the temperature rise of the substrate due to the impact of sputtered atoms determines the average energy. However, this method must take into account the heat of condensation and the various energetic processes occurring in the plasma. Detailed data, however, have been obtained by the time-of-flight method.

Figure 3.14 shows the experiment for the time-of-flight measurement used by Stuart and Wehner (43). The target in a low-pressure dense, dc plasma is pulsed to a fixed negative voltage so that atoms are sputtered from the target as a group. When the group of

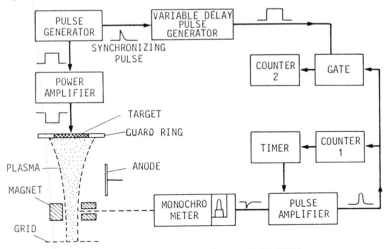

Figure 3.14: Time of flight method (Stuart, Wehner, 1964 (43)).

atoms travels in the plasma the atoms are excited and emit their characteristic spectrum. The energy distribution is determined by the measurement of the time shift of the emitted spectrum.

Typical results for Cu target sputtered by Kr ions are shown in Fig. 3.15 (43). It shows that the kinetic energy of sputtered atoms is in the range of 0 to 40 eV for Cu (110) normally bombarded by 80 to 1200 eV Kr ions. The energy distribution peaks at a few electron volts and more than 90% of the sputtered atoms have energies greater than 1 eV.

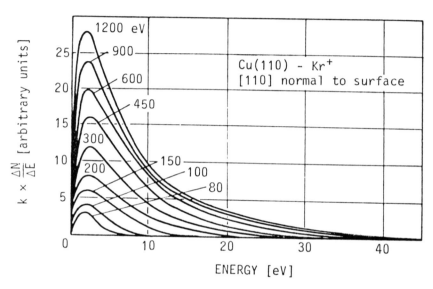

Figure 3.15: Energy distributions of sputtered atoms for various incident ion energy (Stuart, Wehner, 1964 (43)).

The energy of the sputtered atoms is dependent on both the incident ion species and incident bombardment angles. Typical results are shown in Fig. 3.16 and 3.17. Oblique ion incident shifts the energy distribution to higher region. Figure 3.18 shows the variation of the average energy of the sputtered atoms with the incident ion energy. Note that the average energy of the sputtered atoms shows a slight dependence on the incident ion energy. The value will saturate near 10 eV.

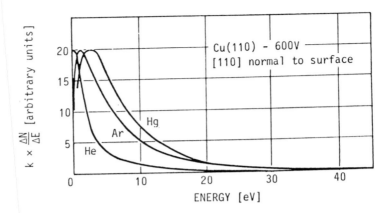

Figure 3.16: Energy distributions of sputtered atoms for various incident ions (Stuart, Wehner, 1964 (43)).

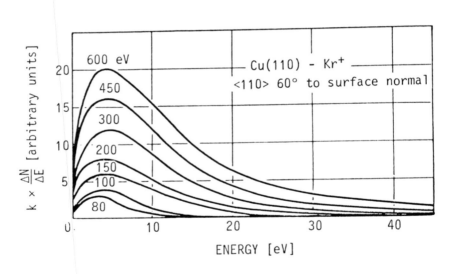

Figure 3.17: Energy distributions of sputtered atoms for various ion energy (Stuart, Wehner, 1964 (43).

These results suggest that when the incident ion energy increases, energy losses in the target also increase such that the energy of incident ions is not effectively transferred to sputtered atoms. For light ion bombardment, ions will penetrate beneath the target surface and their energy loss will increase. This will shift the energy distribution to a lower energy level.

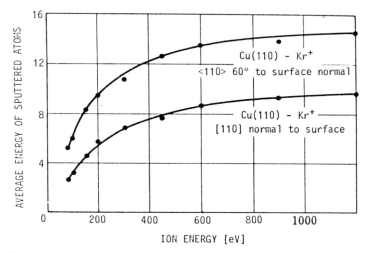

Figure 3.18: Average energy of sputtered atoms (Stuart, Wehner, 1964 (43)).

In the case of sputtered ions their average energy is higher than that of sputtered neutral atoms described above. This experiment is complicated by the presence of a strong electric field at the cathode surface in a plasma-based sputtering measurement. This strong field will tend to return emitted positive ions to the cathode surface. Sputtered ions which escape from the target are limited to very highly energetic sputtered ions. In an ion beam experiment, which has no strong field at the cathode surface, the amount of sputtered ions is on the order of 1% of the sputtered atoms.

The knowledge of the composition and the kinetic properties of sputtered particles is important for the understanding of sputtering mechanism, thin film growth and/or surface science. Extensive studies have continued on the study of the sputtered particles (44) including the postionization by an electron cyclotron resonance (ECR) plasma (45), multiphoton resonance ionization (MPRI) (46), and laser-induced fluorescence spectroscopy (LFS) (47).

Since the vast majority of sputtered particles are emitted as neutrals, most experiments rely on some sort of ionization technique, and subsequent detection of the ions. The most simple technique relies on postionization by an electron beam. However, the postionization probabilities are estimated to approach at best values around 10^{-4}. The postionization by electron impact in a low pressure noble gas plasma excited by an ECR plasma is much more efficient in ionizing the sputtered atoms, as is the MPRI technique. The LFS is also an attractive method for the understanding of the energy distribution of the sputtered particles. The velocity of the sputtered particles can be measured by the Doppler-shift of the emitted light. These experimental arrangements are shown in Fig. 3.19 (45,48).

Typical energy distributions measured by the ECR plasma ionization system are shown in Fig. 3.20 (45).

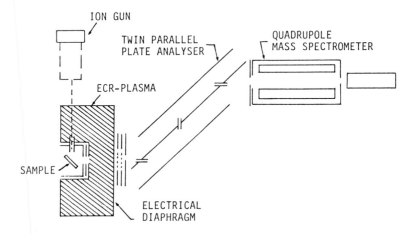

Figure 3.19: Experimental arrangements for the study of sputtered particles postionized by an ECR plasma source (Wucher, Oechsner, 1988 (48).

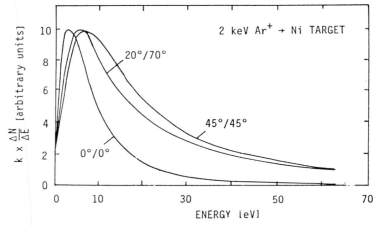

Figure 3.20: Energy distributions of sputtered atoms for various incident/escape angles for Ar+ bombardment of Ni (Oechsner, 1989 (45)).

Experimentally, the E^{-2} dependence at the high energy slope of N(E) has been observed which denotes the validity of isotropic sputtering cascades theory in the solid target developed by M. W. Thompson (49).

The LFS study suggests that under anisotropic conditions such as a low energy region and oblique incidence the isotropic theory is insufficient. The velocity distributions have a "hot" tail compared to the isotropic model (49).

3.2.2.2 Mean Free Path: Most of the sputtered atoms pass through the discharge space in the form of neutral atoms. Their mean free path before they collide to discharge gas molecules is given by

$$\lambda_1 \simeq c_1/\nu_{12}, \tag{3.8}$$

where c_1 is the mean velocity of sputtered atoms and ν_{12} is the mean collision frequency between sputtered atoms and discharge gas molecules. Since the velocity of sputtered particles is much larger than the gas molecules, ν_{12} is given by

$$\nu_{12} \simeq \pi (r_1 + r_2)^2 \bar{c}_1 n_2, \tag{3.9}$$

where r_1 and r_2 are the atomic radius of sputtered atoms and discharge gas molecules, respectively, n_2 is the density of discharge gas. Then the mean free path is simply given by

$$\lambda_1 \simeq 1/\pi (r_1 + r_2)^2 n_2. \tag{3.10}$$

In the case of Ar ion bombardment of a copper target, taking $r_1 = 0.96 \times 10^{-8}$cm, $r_2 = 1.82 \times 10^{-8}$cm, and $n_2 = 3.5 \times 10^{16}$/cm^3(at $0°$C, 1 Torr), λ_1 becomes 11.7×10^{-3}cm. These estimated values are slightly longer than the mean free path between neutral gas molecules at room temperature.

3.3 MECHANISMS OF SPUTTERING

Two theoretical models have been proposed for sputtering:

1. Thermal vaporization theory; the surface of the target is heated enough to be vaporized due to the bombardment of energetic ions.

2. Momentum transfer theory; surface atoms of the target are emitted when kinetic moments of incident particles are transferred to target surface atoms.

The thermal vaporization theory was supported by Hippel in 1926, Sommermeyer in 1935 and Townes in 1944 due to their experimental observations of the Kundsen cosine emission distribution. At that time, the thermal vaporization theory was considered the most important mechanism.

The momentum transfer theory was first proposed by Stark in 1908 and Compton in 1934. The detailed studies by Wehner in 1956, including the observation of spot patterns in single crystal sputtering, suggested that the most important mechanism is not thermal vaporization but the momentum transfer process. At present sputtering is believed to be caused by a collision cascade in the surface layers of a solid.

3.3.1 Sputtering Collisions

The nuclei of the target atoms are screened by electron clouds. The type of collision between an incident particle and the target is determined by incident ion energy and the degree of electron screening (50).

The effects of electron screening are considered as Coulomb collisions. The interaction between two atoms is given by

$$\pm \frac{z_1 z_2 e^2}{r} \exp(-r/a), \tag{3.11}$$

where r is the distance between incident ions and target surface atoms, $z_1 e$ and $z_2 e$ are the nuclear charge of incident ions and target atoms respectively, and a is the screening radius. The a varies with the degree of electron screening. For the Thomas-Fermi potential, $a = c a_h/z_2^{1/3}$, where a_h is the Bohr radius ($= \hbar^2/me^2$), and $c \simeq 1$.

The degree of electron screening expressed by (3.11) and the distance of closest approach determine the type of collisions between incident ions and target atoms. In the case of head-on collisions between these atoms, the distance of closest approach b is given by the relation

$$\frac{z_1 z_2 e^2}{b} = M_1 M_2 v_1^2 / 2(M_1 + M_2), \tag{3.12}$$

where M_1 and M_2 are the mass of the incident ion and the target atoms respectively, and v_1 is the velocity of the incident ions. Combining (3.11) and (3.12) leads to

$$b/a = \frac{2 z_1 z_2 e^2}{M_1 v_1^2} \frac{z_2^{1/3}}{a_h} = z_1 z_2^{4/3} 2 R_h/E, \tag{3.13}$$

where R_h is the Rydberg energy ($e^2/2a_h = 13.54 eV$) and E is the energy of the incident ion energy.

The state of collisions varies with the ratio of b/a. At high energies, i.e., $b/a \ll 1$, the incident ions are scarcely screened by the electron clouds of the target atoms. The incident ions are scattered by target atoms. This is similar to Rutherford scattering. At low energies, i.e., $b/a \gg 1$, the incident ions are screened by the electron clouds of target atoms. The collisions are considered by the classical hard-sphere model.

The sputtering phenomena at moderate energies is believed to be caused by a collision cascade in the surface layers of the target. For higher order collisions near the target surface, the energy of the collision atoms will be much less than the incident recoil energy from the first collision (several hundred eV). From (3.13), this suggests that the approximation $b/a \ll 1$ is valid. The approximation $b/a \gg 1$, however, will not fit for the primary collision, so the whole sputtering process can not be understood by the simple hard-sphere model, where $b/a \gg 1$. Only roughly can sputtering be considered by the simple hard-sphere model.

For $b/a \ll 1$ and $b/\lambda \gg 1$, the collision is governed by the classical Rutherford scattering. A differential cross section $R(\theta)dw$ (dw denotes an elements of solid angle) is expressed by

$$R(\theta) = \frac{b^2}{8} \operatorname{cosec}^4 \theta/2, \qquad \tan \theta/2 = b/2p. \tag{3.14}$$

For b/a<<1 and b/λ <<1, a differential cross section B(θ)dw is given by

$$B(\theta) = \frac{R(\theta)}{[1 + (\lambda/2a)/\sin \theta/2)^2]^2} \quad , \tag{3.15}$$

under the assumption that the collision is considered as screened by a Coulomb scattering event with Born's approximation.

For b/λ >> b/a >> 1 the collision is considered by the elastic collision theory. The collision is isotropic and the total scattering cross section is expressed by πR^2. From Eq. (3.11) we have R = a $\log(z_1 z_2 e^2/RE)$ for an incident ion energy, E. For a small E the R is approximately the atomic radius of the incident particles.

For b/λ < b/a, the cross section is not calculated simply by Born's approximation except in the case b/λ < < (b/a)$^{1/2}$. For b/λ < < (b/a)$^{1/2}$ the collision is isotropic.

3.3.2 Sputtering

A sputtering event is initiated by the first collision between incident ions and target surface atoms followed by the second and the third collisions between the target surface atoms. The displacement of target surface atoms will eventually be more isotropic due to successive collisions and atoms may finally escape from the surface. Figure 3.21 shows the features of sputtering collision in the target surface.

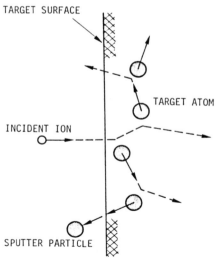

TARGET SURFACE

TARGET ATOM

INCIDENT ION

SPUTTER PARTICLE

Figure 3.21: Incident ions and the sputtered particles.

The sputtering process is considered in detail for the following three different energy region of the incident ions (51).

1.   threshold region ($<$ 100eV)

2.   low energy region (0.1 keV)

3.   high energy region ($10 \simeq 60$ keV)

Detailed considerations have been studied with the aid of computer simulation (52,53).

According to the elastic-collision theory, the maximum possible energy transferred in the first collision T_m is given by

$$T_m = \frac{4M_1 M_2}{(M_1 + M_2)^2} E, \qquad (3.16)$$

where M_1 and M_2 are the masses of the incident ions and target atoms respectively, and E is the energy of incident ions. In the first order of approximation the sputter yield S is proportional to the T_m, the sputter yield of a given target material bombarded with different element is given by

$$S = k \frac{1}{\lambda(E) \cos \theta} \frac{M_1 M_2}{(M_1 + M_2)^2} E, \qquad (3.17)$$

where k is a constant which includes different target material constants, λ is the mean free path for elastic collisions near the target surface, and θ is the angle between the normal on the target surface and the direction of incidence ions. The mean free path is given by

$$\lambda = \frac{1}{\pi R^2 n_0}, \qquad (3.18)$$

where n_0 is the number of lattice atoms per unit volume, and R is the collision radius.

The collision radius R for the rigid sphere model can be calculated for a screened potential as

$$R = C \frac{a_0}{(z_1^{2/3} + z_2^{2/3})^{1/2}} \ln \frac{z_1 z_2 e^2}{\varepsilon_0 RE'}, \qquad (3.19)$$

where $E' = M_1 E / (M_1 + M_2)$, C is a constant, a_0 is the radius of the hydrogen atom ($= 0.57 \times 10^{-8}$cm), e the elementary charge, ε_0 the dielectric constant in the vacuum, $Z_1 e$ and $Z_2 e$ are the nuclear charges for M_1 and M_2 respectively. The relation (3.17) gives qualitative information about the sputter yield.

Rol et al have suggested that by putting $k = 1.67 \times 10^{-11}$ m / eV, C = 1, the sputter yield measured for the ion bombardment of copper target with 5 to 20keV Ar of N ions fit the theoretical relationship (3.17) (54). Almen et al. have shown that the constant k is expressed by the experimental relationship

$$k = a \exp(\frac{-b\sqrt{M_1}}{M_1 + M_2} E_B),$$

and found the sputter yield as

$$S = 4.24 \times 10^{-8} n_0 R^2 E \frac{M_1 M_2}{(M_1 + M_2)^2} \exp(-10.4 \frac{\sqrt{M_1}}{M_1 + M_2} E_B),$$

(3.20)

where E_B is the binding energy of the target materials (55).

Goldman and Simon have shown the theoretical model of sputtering of copper with 500keV deuterons (56). Rutherford scattering will take place in the collisions between the deuterons and target copper atoms. In these collisions the mean free path of the deuteron is of the order 10^{-4} cm. The energy for the displacement of target atoms from their normal lattice is about 25eV. The average energy of the target recoil atoms after the deuteron bombardment at the first collision is estimated to be 200 eV.* Thus the collisions after the first collision are considered by the simple hard-sphere model. The mean free path after the first collision is of the order of 10^{-7} cm. The diffusion of these knock-on atoms is treated by simple theory and the sputter yield is given by the relationship

$$S \propto \frac{M_1}{M_2} \frac{\ln E}{E} \frac{1}{\cos \theta}.$$

(3.21)

Almen et al. have experimentally shown the validity of the relationship (3.21) (55).

Sigmund has studied the theory of sputtering in detail. He assumed that sputtering of the target by energetic ions or recoil atoms results from cascades of atomic collisions. The sputtering yield is calculated under the assumption of random slowing down in an infinite medium. The theoretical formula was compared with the experimental results given by Wehner (11,57).

---

* When incident atoms with energy E, mass M_1 elastically collides with the target atom with mass M_2 the mean free path of the incident atom, λ_i is expressed by

$$\lambda_i = (n_0 \sigma_d)^{-1}, \quad \sigma_d = \pi \frac{M_1 z_1^2 z_2^2 e^4}{M_2 E E_d},$$

where n_0 is the atomic density of the target, and E_d is the energy for the displacement from the normal lattice. The average energy of the recoil atoms T_d is expressed by

$$\overline{T}_d = E_d \ln \frac{4 M_1 M_2 E}{(M_1 + M_2)^2 E_d}.$$

In theory the collision is considered to be the hard sphere model. The sputter yield is expressed by the relationships

(i) < 1 keV

$$S(E) = (3/4\pi^2)\alpha T_m/U_0, \tag{3.22}$$

(ii) $1 \simeq 10$ keV

$$S(E) = 0.420\alpha S_n(E)/U_0, \tag{3.23}$$

where

$$T_m = \frac{4M_1M_2}{(M_1 + M_2)^2} E,$$

$$U_0 = \text{heat of sublimation},$$

$$\alpha; \text{ function of } M_2/M_1$$

$$S_n(E) = 4\pi z_1 z_2 e^2 a_{12}[M_1/(M_1 + M_2)]s_n(\varepsilon)$$

$$\varepsilon = \frac{M_2 E/(M_1 + M_2)}{z_1 z_2 e^2 a_{12}}.$$

$$a_{12} = 0.8853 a_0(z_1^{2/3} + z_2^{2/3})^{-1/2}$$

$$s_n(\varepsilon); \text{ reduced energy stopping cross section for T-F interaction}$$

The value α and $s_n(\varepsilon)$ are shown in Fig. 3.22 and Table 3.7. A Born-Mayer potential $(V(r) = Ae^{-r/a}, a = 0.219\text{Å}, A = 52 (Z_1Z_2)^{3/4} \text{ eV})$ is applied in the low energy region, (3.22), while the Thomas-Fermi potential is applied in the high energy region, (3.23).

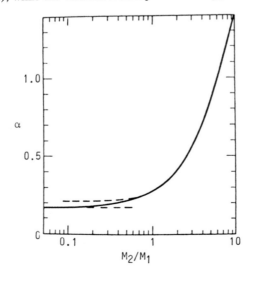

Figure 3.22 The value, α due to (Sigmund, 1968 (7)).

ε	$s_n(\varepsilon)$	ε	$s_n(\varepsilon)$
0.002	0.120	0.4	0.405
0.004	0.154	1.0	0.356
0.01	0.211	2.0	0.291
0.02	0.261	4.0	0.214
0.04	0.311	10	0.128
0.1	0.373	20	0.0813
0.2	0.403	40	0.0493

Table 3.7: The values, $s_n(\varepsilon)$ (Sigmund, 1969 (11)).

Figure 3.23 shows the comparison of measured and theoretical yield values at energy below 1 keV. The agreement is quite good. The periodicity relates to the U_0. Figure 3.24 shows sputter yields of polycrystalline copper for Kr ions. At the ion energy below 10 keV, (3.22) and (3.23) agree well with the experiments.

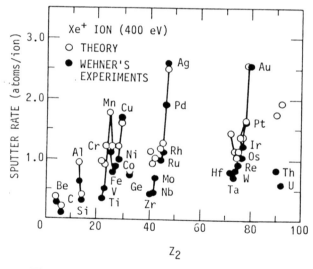

Figure 3.23 Sputter yields for the various materials compared with Sigmund's theory (Sigmund, 1969 (11)).

The linear cascade theory developed by Sigmund can satisfactorily account for the sputtering behavior of amorphous or polycrystalline elemental targets. A possible exception to this theory is the case of anisotropic collisions including the single target, the low energy region, and/or the oblique incidence.

More recently, theoretical studies on sputtering have been developed by many workers. Harrison et al. have shown the computer simulation of the emission by sputtering of clusters (53). They suggest that the emission of clusters of two and three atoms are high-probability events in sputtering of (100) copper by Ar ions in the energy range from 0.5 to 5.0 keV.

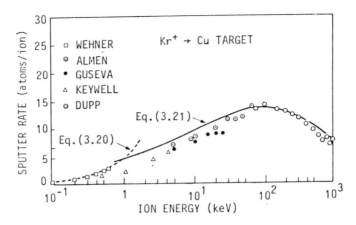

Figure 3.24 Sputter yields calculated by Sigmund (Sigmund (1969) 7).

In the case of a liquid target, it is known that threshold values are smaller than values of a solid target due to the weaker binding energy in the liquid. In the high energy region, the sputter yields decrease significantly because of the increase of energy losses in the liquid target. For example, when sputtering Al by Hg ions, the yield decreases by 30% when the Al in molten.

Recent developments on the surface analysis, such as AES, ESCA and SIMS help examine the sputtering features of an alloy. Altered layers are formed on the alloy target and enhanced surface and near surface diffusion are observed during sputtering. This sometimes causes difficulty in the controlled deposition of alloy films by direct sputtering of the alloy target (58).

3.4 REFERENCES

1. Zalm, P.C., in Handbook of Ion Beam Processing Technology (J.J. Cuomo, S. M. Rossnagel and H. R. Kaufman,eds.) p.78. Noyes Publications, New Jersey (1989)

2. Sartwell, B.D., J. Appl. Phys., 50: 78 (1979).

3. Kirschner, J., Etzkorn, H.W., Appl. Phys. A 29: 133 (1982).

4. Hull, A.W., Winter, H.F., Phys. Rev., 21: 211 (1923).

5. Holland, L., Vacuum Deposition of Thin Films, p. 408, London, Chapman & Hall Ltd., (1961).

6. Wehner, G.K., Phys. Rev., 102: 690 (1956), ibid.,108: 35 (1957), ibid., 112: 1120 (1958.

7. Stuart, R.V. and Wehner, G.K., J. Appl. Phys., 33: 2345 (1962).

8. Harrison, D.E., Magnuson, G.D., Phys. Rev., 122: 1421 (1961).

9. Yamamura, Y., and Bohdasky, J., Vacuum, 35: 561 (1985).

10. Wehner, G.K., Phys. Rev., 108: 35 (1957).

11. Sigmund. P., <u>Phys. Rev.,</u> 184: 383 (1969).

12. Wehner, G.K., <u>Advances in Electronics and Electron Physics, Vol. VII.,</u> (Academic Press, Inc., New York, 1955)Vol.VII.

13. Massey, H.S.W., Burhop, E.H.A., <u>Electronics and Ionic Impact Phenomena</u> New York, Oxford Univ. Press (1952).

14. Laegreid, N., Wehner, G.K., Meckel, B., <u>J. Appl.Phys.,</u> 30: 347 (1959).

15. Laegreid, N., Wehner, G.K., <u>J. Appl. Phys.,</u> 32: 365 (1961).

16. Brown, S.C., <u>Basic Data of Plasma Physics</u> Cambridge, MA, MIT Press, p.230 (1959).

17. Henschke, E.B., <u>Phys. Rev.,</u> 106: 737 (1957).

18. Wehner, G.K., <u>J. Appl. Phys.,</u> 30: 1762 (1959).

19. Wehner, G.K., Rosenberg, D.L., <u>J. Appl. Phys.,</u> 31: 177 (1960).

20. Hoffmann, T., Dodds, H.L, Robinson, M.T., Holmes, D.K., <u>Nucl. Sci. Eng.</u> 68: 204 (1978).

21. Anderson, H.H., Bay, H.L., in <u>Sputtering by Particle Bombardment I,</u> p. 202, (R. Behrisch, ed.), Springer Verlag, Berlin (1981).

22. Okutani, T., Shikata, M., Ichimura, S., Shimizu, R., <u>J. Appl.Phys.,</u> 51: 2884 (1980).

23. Seeliger, R., Sommermeyer, K., <u>Z. Physik,</u> 93: 692 (1935).

24. Maissel, L.I., Glang, R., (eds.), <u>Handbook of Thin Film Technology</u> p. 3-25, New York, McGraw Hill Inc. (1970).

25. E. Roosendaal, H.E., in <u>Sputtering by Particle Bombardment Vol. I,</u> (R. Behrish, ed.) p. 217 Springer Verlag, Berlin (1983).

26. Jackson, D.P., <u>Radiat. Eff.</u> 18: 185 (1973).

27. Roosendaal, H.E., in <u>Sputtering by Particle Bombardment Vol. I,</u> (R. Behrish, ed.) p. 224, Springer Verlag, Berlin (1983).

28. Fluit, J.M., Rol, P.K., Kristemaker, J., <u>J. Appl. Phys.,</u> 34: 3267 (1963).

29. Onderdelinden, D., <u>Can.J.Phys.,</u> 46: 739 (1968).

30. Elich, J. J. Ph., Roosendaal, H.E., Onderlinden, D., <u>Radiat. Eff.</u> 14: 93 (1972).

31. Flur, B.L., Riseman, J., <u>J. Appl. Phys.,</u> 35: 344 (1964).

32. Patterson, W.L., Shirn, G.A., <u>J. Vac. Sci. Technol.,</u> 4: 343 (1967).

33. Liau, Z.L., Brown, W.L., Homer, R., Poate, J.M., <u>Appl. Phys. Lett.,</u> 30: 626 (1977).

34. Haff, P.K., Switkowski, Z.E., <u>Appl. Phys. Lett.,</u> 29: 549 (1976).

35. Tarng, M.L., Wehner, G.K., <u>J. Appl. Phys.,</u> 42: 2449 (1971).

36. Wilson, R.L., Terry, L.E., <u>J. Vac. Sci. Technol.,</u> 13: 157 (1976).

37. G.S. Anderson, <u>J. Appl. Phys</u> 40: 2884 (1969).

38. Maissel, L.I., Glang, R. (eds.) <u>Handbook of Thin Film Technology</u> p. 3-23, New York, McGraw Hill (1970).

39. Guenthershulze, A., <u>Z. Physik.,</u> 119: 79 (1942).

40. Mayer, H., <u>Phil. Mag.,</u> 16: 594 (1933).

41. Sporn, H., Z. Physik., 112: 278 (1939).

42. Wehner, G.K., Phys. Rev., 114: 1270 (1959).

43. Stuart, R.V., Wehner, G.K., J. Appl. Phys., 35: 1819 (1964).

44. Anderson, H.H., Bay, H.L., in Sputtering by Particle Bombardment Vol. I, (R. Behrisch, ed.) p. 157, Springer Verlag, Berlin (1983).

45. Oechsner, H., in Handbook of Ion Beam Processing Technology, p. 145 (J.J. Cuomo, S. M. Rossnagel, and H.R. Kaufman, eds.) Noyes Publications, New Jersey (1989).

46. Pappas, D.L., Winograd, N., and Kimok, F.M., in Handbook of Ion Beam Processing Technology, p. 128 (J.J. Cuomo, S. M. Rossnagel, and H.R. Kaufman, eds.) Noyes Publications, New Jersey (1989).

47. Calaway, W.F., Young, C.E., Pellin, M.J., and Gruen, D.M., in Handbook of Ion Beam Processing p. 112 (J.J. Cuomo, S. M. Rossnagel, and H.R. Kaufman, eds.) Noyes Publications, New Jersey (1989).

48. Wucher A., and Oechsner, H., Surface Sci. 199: 567 (1988).

49. Tompson, N.W., Phil. Mag., 18: 337 (1968).

50. Kotani, M., Banno, Y., Fukada, E., Radiophysics, Tokyo, Iwanami, p. 18 (1959).

51. Wilson, R.G., Brown, G.R., Ion Beams, p. 322, New York, John Wiley & Sons (1973).

52. Harrison, D.E., Levy, N.S., Johnson, J.P., Effron, H.M., J. Appl. Phys., 39: 3742 (1968).

53. Harrison, D.E., Delaplain, C.B., J. Appl. Phys., 47: 2252 (1976).

54. Rol, P.K., Fluit, J.M., Kistemaker, J., Physica., 26: 1009 (1960).

55. Almen, O., Bruce, G., Nucl. Instr. Methods, 11: 257, 279 (1961). 279.

56. Goldman, D.T., Simon, A., Phys. Rev., 111: 383 (1958).

57. Rosenberg, D.L., Wehner, G.K., J. Appl. Phys., 33: 1842 (1962).

58. Saeki, N., Shimizu, R., Jpn. J. Appl. Phys., 17: 59 (1978).

4

SPUTTERING SYSTEMS

Technological advances in the sputtering process in the last 10 years have allowed sputtering systems to be widely used industrially as an important alternative to conventional thermal evaporation systems and chemical vapor deposition systems for making LSI.

The understanding of glow discharges is important in order to master the sputter deposition system, since virtually all of the energetic incident particles originate in the plasma.

In this chapter, the basic concepts of the glow discharge are given, after which the construction and operation of the sputtering deposition system are described.

4.1 DISCHARGE IN A GAS

4.1.1 Cold-cathode Discharge

In a diode gas discharge tube the minimum voltage which initiates the discharge, i.e. the spark or breakdown voltage V_s, is given by

$$V_s = a \frac{pl}{\log pl + b},$$

(4.1)

where p is the gas pressure, l is the electrode separation, a and b are constant. The relationship between spark voltage V_s and the gas pressure p is called Paschen's law. A typical experimental result is shown in Fig. 4.1, which shows the existence of minimum spark voltage at gas pressure P_m (1).

In a conventional sputtering system, the gas pressure is kept below the P_s. To initiate the discharge, gas pressure p_s under a given electrode separation l is expressed by

$$p_s \simeq \lambda_0/l,$$

(4.2)

where λ_0 is mean the free path of electrons in the discharge gas, which is given by $\lambda_0 = 1/P_c$, where P_c denotes the elastic-collision cross section between electrons and gas atoms. The P_c in Ne, Ar, Kr and Xe is shown in Fig. 4.2 (2). Taking $P_c = 20 \text{cm}^{-1}\text{Torr}^{-1}$ in Ar, at 100eV, λ becomes 0.05 cm. Putting l = 10 cm, p_m becomes 5 mTorr. This suggests the gas pressure should be higher than 5 mTorr for initiating the breakdown, and forming the discharge.

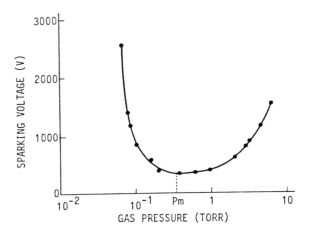

Figure 4.1: Sparking voltage vs. gas pressure measured for Cu electrodes in air (electrode separation, 5mm)

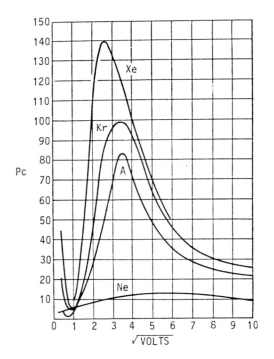

Figure 4.2: Probability of collision in Ne, Ar, Kr, (Brown, 1959 (2)).

In a diode discharge tube, when the discharge is initiated in a low-pressure gas with a high impedance dc power supply, the mode of discharge varies with the discharge current. Figure 4.3 shows the discharge mode (3).

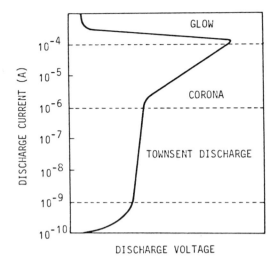

Figure 4.3: A classification of gas discharge for various discharge currents (Brown, 1959 (2)).

At a discharge current below 0.1mA/cm², the Townsend discharge appears. A small, non-self-sustained continuous current is maintained in the Townsend discharge. Positive ions are created close to the anode due to a multiple avalanche, traverse the whole gap and produce a uniform space charge. However, the effects of the space charge are not dominant.

At the discharge current above 0.1mA/cm², a luminous glow is observed near the cathode. This region is called a glow discharge and the effects of the space charge are dominant.

At the low current end of the glow discharge region, the glow partially covers the cathode surface. This mode is called normal glow discharge. When the current density increases, the glow covers the whole cathode. This mode is called the abnormal glow discharge. In the abnormal glow discharge, if the cathode is not cooled and the current increases above 0.1A/cm², thermionic electrons are emitted. This causes a transition into the arc discharge.

The basic ionization process in a gas discharge is as follows. When the electrons collide with gas molecules, the latter are ionized and positive ions appear. The energy of the electrons should be higher than the ionization energy of the gas molecules. At the beginning of the discharge, the primary electrons from the cathode are accelerated by the electric field near the cathode and an energy in excess of the ionization energy of the gas molecules. These energetic electrons collide with the gas molecules and generate positive ions before they travel to the anode. The positive ions bombard the cathode surface which result in the generation of secondary electrons from the cathode surface. The secondary electrons increase the ionization of the gas molecules and generate a self-sustained discharge.

When the discharge current is below 10^{-9} A, the secondary electrons are insufficient in number to cause enough ionization to produce a self-sustained discharge, as shown in

Fig. 4.3.   The self-sustained discharge appears when the discharge current is above $\approx 10^{-9}$A. This discharge is characterized by positive space charge effects in the edge regions of the plasma.

When a single primary electron collides with m gas molecules and creates m numbers of electrons and ions, the self-sustained discharge will appear if

$$m\gamma = 1, \tag{4.3}$$

where γ denotes the number of secondary electrons per incident ion impact on the cathode surface (also known as the secondary electron coefficient).

The rate at which electrons are created in the discharge due to collisions with the secondaries is given as: $dn_e/dx = \alpha n_e$, where n_e denotes the number of electrons and α the primary ionization coefficient. Thus the single primary electron creates $\exp(\alpha l)$ secondary electron during the travel of electrode spacing. The multiplication coefficient of the primary electron m is given by $m = [\exp(\alpha l) - 1]$.

Since $m = 1/\gamma$,

$$\alpha l = \ln(1 + 1/\gamma). \tag{4.4}$$

The α is empirically expressed by the following relationship

$$\alpha/p = A \exp[-B/(E/p)], \tag{4.5}$$

where E denotes the electric field in the discharge region, p the pressure of the discharge gas, and A and B are constant. Assuming $E = V_s/l$, Pashen's law described in Fig. 4.1 is given by

$$V_s = \frac{B(pl)}{\ln(pl) + \ln[A/[\ln(1 + 1/\gamma)]]}. \tag{4.6}$$

Table 4.1 shows the values A and B with ionization energy.

In a glow discharge, the potential distribution between electrodes is nonuniform due to the presence of the charge as shown in Fig. 4.4. There is a voltage drop V_c near the cathode, which is known as the cathode fall. The cathode fall region corresponds to a so-called cathode dark space or Crookes dark space. The spacing d corresponds to the region through which the electron gains at least the ionization energy of gas molecules. The cathode fall satisfies the relationship $\exp \int_0^d \alpha dx = 1 + 1/\gamma$, where α is the ionization coefficient, and γ the secondary electron emission coefficient.

GAS	A ION PAIRS cm X TORR	B V cm X TORR	E/p V cm X TORR	V_i (V)
H_2	5	130	150~ 600	15.4
N_2	12	342	100~ 600	15.5
O_2	—	—	—	12.2
CO_2	20	466	500~1,000	13.7
air	15	365	100~ 800	—
H_2O	13	290	150~1,000	12.6
HCl	25	380	200~1,000	—
He	3	34(25)	20~ 150(3~10)	24.5
Ne	4	100	100~ 400	21.5
A	14	180	100~ 600	15.7
Kr	17	240	100~1,000	14
Xe	26	350	200~ 800	12.1
Hg	20	370	200~ 600	10.4

Table 4.1: Ionization energy, and constants A and B for various gases (Brown, 1959 (2)).

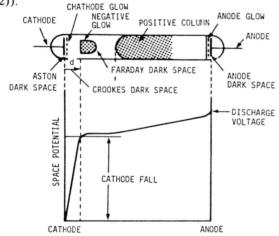

Figure 4.4: Features of a glow discharge.

The electrical field E(x) in Crookes dark space is empirically shown by the relationship E=k(d-x). The spatial distribution of the potential V(x) is given by $V(x) = \int_0^x E dx = \int_0^x k(d - x)dx$. Since V(x) becomes the cathode fall, V_c at d = x, the V(x) and E(x) are expressed by

$$V(x) = V_c x(2d - x)/d^2, \qquad (4.7)$$

$$E(x) = 2V_c(d - x)/d^2. \qquad (4.8)$$

The space charge accumulated in Crookes dark space is estimated by Poisson's relationship $d^2V/dx^2 = -\rho/\varepsilon_0$. The space charge ρ is given by the relationship

$$\rho = 2\varepsilon_0 V_c/d^2. \tag{4.9}$$

In the cathode fall region the current flow i+ is governed by the space charge limited current (SCLC), since the space charge given by the relationship (4.9) appears. The ions are accelerated by the potential $V(x)$ in the cathode fall region. The velocity of the ions V_+ is given by $V_+ = \sqrt{2eV/M}$, where e and M are the charge and mass of the ions. Since $j_- = \rho v_+$, $d^2V/dx^2 = (j_+/\varepsilon_0)\sqrt{M/2eV}$. Then, the current density j_+ is given by

$$j_+ = 4/9\varepsilon_0(2e/M)^{1/2} \frac{V_c^{3/2}}{d^2}, \tag{4.10}$$

under the condition $V=0$ at $x=0$.

The glow discharge is maintained by secondary electrons produced at the cathode by positive ion bombardment. The current density at cathode is given by

$$j_s = j_+(1 + \gamma). \tag{4.11}$$

In glow discharge sputtering, the energy of the incident ions is close to the cathode fall potential. Since the anode fall, which is 10 to 20 V, and the potential drop across the positive column are much smaller than the cathode fall, the incident ion energy is roughly equal to the discharge potential.

The negative glow results from the excitation of the gas atoms by inelastic collisions between the energetic electrons and gas molecules. The abnormal glow discharge is used for sputtering systems and most processing plasmas. In this case, the energy of incident ions is nearly equal to the discharge voltage, since $V_s = V_c$. Note that V_s and j_s cannot be changed independently. The V_c also varies with j_s/p^2 as indicated in Fig. 4.5 (3). The minimum V_c is obtained at normal glow discharge. In the sputtering system, the V_c is 500 to 1,000V which is characteristic of the abnormal discharge. It is understood that lowering the gas pressure causes a decrease of the current density under constant V_c. In order to keep the current density constant the V_c should be increased.

In glow discharge, the values (pd) are essentially constant where p is the discharge gas pressure, and d the width of the cathode dark space. However, experimental values vary with discharge parameters. Table 4.2 shows typical experimental values of (pd). The values (pd) in normal glow discharges are slightly larger than values at abnormal glow discharges.

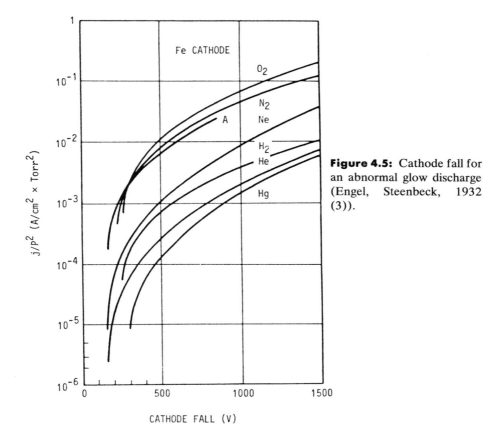

Figure 4.5: Cathode fall for an abnormal glow discharge (Engel, Steenbeck, 1932 (3)).

A superposition of magnetic field or high frequency electrical field somewhat extends the controllable range of discharge parameters. In high frequency discharges, the electrons in the discharge region will have cyclic motions and the number of the collisions will increase between the electrons and the gas molecules. This increases the efficiency of the gas discharge. The effects of the alternating electric field of the discharge, or spark, voltage are expressed by the modified Paschen's law

$$V_s = f(pl, p/\omega), \tag{4.12}$$

where ω is the frequency of the alternating electric field.

(pd) in (Torr x cm)

GAS	NORMAL GLOW	ABNORMAL GLOW DISCHARGE		
		$V_c=500$ V	1000 V	2000 V
H_2	0.9	0.46	0.25	0.2
He	1.3	0.8	0.6	0.5
N_2	0.4	0.17	0.1	0.1
Hg	0.3	0.2	0.12	—
O_2	0.3	0.1	0.07	0.07
A	0.25	0.07	—	—

Table 4.2: Observed (pd) values in (Torr cm.) for an iron cathode (Engel and Steenbeck, 1932 (3)).

4.1.2 Discharge in a Magnetic Field

4.1.2.1 Spark Voltage in a Magnetic Field: In the presence of a magnetic field, electrons in a gas discharge tube show orbital motion around the magnetic lines of force as shown in Fig.4.6. In the magnetic field strength B, the radius of the orbital motion is $r = mv/eB$, where e, m and v are the electron charge, mass, and velocity, respectively. When the magnetic field is longitudinally superposed on the gas discharge, the electrons in the discharge are twined around the magnetic line of force. This reduces the losses of electrons in the discharge region and increases the discharge current density. However, strong additional effects of the magnetic field are observed in the presence of a transverse magnetic field.

When the transverse magnetic field B is superposed on the electric field E, the electron shows cycloidal motion with an angular velocity $\omega = eB/M$ and center of orbit drift in the direction of ExB with the velocity of E/B as shown in Fig. 4.7. These electron motions increase the collision probability between electrons and molecules. This enables one to lower the gas pressure to as low as 10^{-5} Torr.

Figure 4.8 shows the effects of the transverse magnetic field on the spark voltage measured in coaxial cylindrical electrodes, in which the magnetic field is applied in an axial direction (1). It shows the effects of the magnetic field observed at the magnetic field above a critical value B_c, which is the so-called "cut-off field". In the magnetic field below B_c the primary electrons from the vicinity of the cathode will reach the anode without performing the cycloidal motion between the electrode since the radius of the cyclotron motion is greater than the spacing of the electrodes.

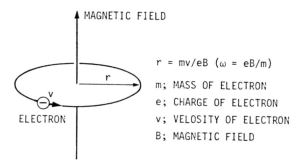

Figure 4.6: Cycloidal motion of electrons in a magnetic field.

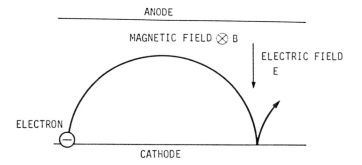

Figure 4.7: Electron trajectory in a crossed electromagnetic field.

The cut-off field B_c is expressed by the following relationships:

$$B_c \simeq (2mV_c/e)^{1/2}\, \frac{2r_2}{r_2^2 - r_1^2} \quad \text{(cylindrical electrode)} \; ,$$

$$\tag{4.13}$$

$$B_c = (2mV_c/e)^{1/2} 1/L \quad \text{(planar electrode)},$$

where V_c is the applied voltage between the electrodes, r_2 and r_1 are the radius of outer and inner electrodes, and L is the electrode spacing.

The abrupt decrease of the discharge voltage in the magnetic field above B_c results from the increase of the collisions between the primary electrons and neutral gas molecules. In a significantly higher magnetic field, the sparking (discharge) voltage increases with the magnetic field. The increase of the voltage in the strong magnetic field is induced by the increase of electron energy losses through successive collisions.

The effects of the transverse magnetic field are qualitatively considered to be an effective pressure, P_e given by

$$p_e/p \approx [1 + (\omega\tau)^2]^{1/2}, \tag{4.14}$$

where ω is the cyclotron frequency of the electron and τ is the mean free time of the electron (4). Since $\omega = eB/M$, and $\tau = \lambda_0/p[2(e/m)V_0]^{1/2}$, $\omega\tau$ becomes

$$\omega\tau \approx \lambda_0 B(e/m)^{1/2}/\sqrt{2}\ pV_0^{1/2}, \tag{4.15}$$

where λ_0 is the mean free path of the electron at 1 Torr, B is the strength of the magnetic field, (e/m) is the specific charge of the electron, and V_0 is the acceleration voltage for the electron. Taking B = 100 G (0.01 T), p = 1×10^{-5}Torr (Ar), V_0 = 100V and λ_0 = 0.05cm, we obtain $\omega\tau = 5\times10^3$, and p_e = 0.05Torr.

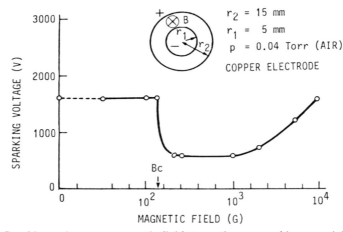

Figure 4.8: Sparking voltage vs. magnetic field strength measured in a coaxial cylindrical electrodes with an axial magnetic field.

Theoretically the spark voltage in the crossed electric and magnetic field is estimated by the change of ionization coefficients α and γ with the superposition of the crossed field. Table 4.3 shows the theoretical formulas of the ionization coefficient α in the crossed field. The effect of the crossed field on γ is not well understood. In general the secondary electrons released from cathode surface due to the effect will be partially recaptured at the cathode surface since the cross field bends their trajectory. This may reduce the effective secondary electron coefficient.

REFERENCES	RELATIONS
Kugler *et al.*	$\alpha = \dfrac{3}{4}\dfrac{eB^2}{mE} : V_E = \dfrac{16}{3\pi}\dfrac{mE^2}{eB^3\lambda}$
Haefer. Somerville	$\alpha = \dfrac{3}{4}\dfrac{eB^2}{mE}\left(1 - \dfrac{B^2 V_i^2}{4E^2}\right)^{\frac{1}{2}}$
Redhead	$\alpha = \dfrac{1}{g}\dfrac{3}{4}\dfrac{eB^2}{mE}\left(1 + \dfrac{B^2 V_i^2}{8E^2}\right)^{-1} : g \approx 3$
Somerville	$\alpha = \dfrac{3}{4}AL\dfrac{eB^2}{mE}\exp\left(-\dfrac{3}{4}bL\dfrac{eB^2}{mE}\right)$
Blevin and Haydon	$\alpha = Ap\left\{1 + \dfrac{B^2 e^2 L^2}{p^2 m^2 u^2}\right\}^{\frac{1}{2}}\exp\left(-\dfrac{bp}{E}\left\{1 + \dfrac{B^2 e^2 L^2}{p^2 m^2 u^2}\right\}^{\frac{1}{2}}\right)$

λ : ELECTRON MEAN FREEPATH
L : λ AT 1 TORR
u : ELECTRON VELOSITY

Table 4.3: Ionization coefficient, α, in a crossed electric and magnetic field (Wasa, Ph.D. Thesis (1961)).

4.1.2.2 Glow Discharge in a Magnetic Field: Under the superposition of a magnetic field crossed to an electric field, the width of the Crookes dark space decreases. This suggests that the superposition of the magnetic field equivalently increases the gas pressure in the discharge region as shown in the relationship (4.14).

We will now consider the glow discharge with parallel electrodes shown in Fig. 4.9. The motion of a primary electron is determined by the equations,

$$m\,\frac{d^2x}{dt^2} = eE - Be\,\frac{dy}{dt},$$

$$m\,\frac{d^2y}{dt^2} = Be\,\frac{dx}{dt},$$

(4.16)

where $E = 2V_c(d - x)/d^2$. From Eqns. (4.16), the equation of electron motions becomes

$$m\,\frac{d^2x}{dt^2} + (ce + B^2e^2/m)x = cde,$$

(4.17)

where $c = 2V_c^2/d^2$. On the assumption that the electron starts from the cathode with zero initial velocity, the electron moves in a cycloidal path. The equation of electron motion for x-direction is given by

$$x = \frac{cd}{c + B^2 e/m}(1 - \cos \omega t),$$

(4.18)

where $\omega = (ce + B^2 e^2/m)/m$.

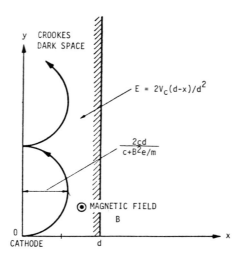

Figure 4.9: Electron motion in a Crookes dark space.

The maximum displacement in the direction of the electric field D is given by

$$D = \frac{2cd}{c + B^2 e/m} .$$

(4.19)

The magnetic field affects the discharge when

$$D = \frac{2cd}{c + B^2 e/m} < d,$$

(4.20)

$$\text{or} \quad B^2 e/m > 2V_c/d^2.$$

The rf-discharge with the transverse magnetic field is often used for the thin film processing. From equations (4.16), the equations of electron motion in the rf discharge are given by

$$m \frac{d^2 x}{dt^2} = eE_0 \sin \omega t - Be \frac{dy}{dt} ,$$

(4.21)

$$m \frac{d^2 y}{dt^2} = Be \frac{dx}{dt} ,$$

when we substitute E by $E_0 \sin \omega t$. From equation (4.21), the electron velocity in the direction of electric field dx/dt is given by

$$\frac{dx}{dt} = v_0 \sin \omega_H t + u_0 \cos \omega_H t - \frac{eE_0}{m} \frac{\omega}{\omega_H^2 - \omega^2} \cos \omega_H t + \frac{eE_0}{m} \frac{\omega}{\omega_H^2 - \omega^2} \cos \omega t,$$

(4.22)

where u_0 and v_0 denote the initial velocity of electrons in x and y directions, respectively, and $\omega_H = eB/M$, the electron cyclotron frequency. The energy ε transferred to the electrons during the cycloidal motion of the electrons is given by

$$\varepsilon = \frac{eE_0^2}{8m} \left[\frac{1}{(\omega - \omega_H)^2} + \frac{1}{(\omega + \omega_H)^2} + \frac{2 \cos 2\omega t}{\omega_H^2 - \omega^2} \right]. \tag{4.23}$$

Equations (4.22) and (4.23) show that electrons will effectively receive energy at the cyclotron frequency $\omega = \omega_H$. The discharge voltage exhibits a minimum point at $\omega = \omega_H$.

4.1.2.3. Glow Discharge Modes in the Transverse Magnetic Field: In the presence of a transverse magnetic field, there appears to be two different modes of glow discharge in a dc cold cathode discharge tube. One is a positive space-charge-dominated mode (PSC mode) which appears in a weak magnetic field. The other is a negative space-charge-dominated mode (NSC mode) which appears in a strong magnetic field. The appearance of the different modes is interpreted in the following manner: in the presence of a magnetic field the path of the electrons is deflected. The electrons perform a cycloidal motion in the plane perpendicular to the magnetic field between the electrodes. As a result the electron radial velocity is reduced, and decreases with the increase of the magnetic field. The ion radial velocity is only marginally changed by the magnetic field. When the magnetic field is so strong that the electron velocity is smaller than the ion velocity, the NSC mode appears. In this mode, the cathode fall is very small and inversely the anode fall is very large. The electron gains sufficient energy to cause ionization of neutral molecules in the anode fall. When the magnetic field is so weak that the electron velocity is larger than the ion velocity, the PSC mode appears. In this mode the anode fall is very small, and inversely there is a large cathode fall by which the electron gain energy to cause ionization. Typical distributions are shown in Fig. 4.10 (5). The transition between the two modes will be observed when $\mu_e/\mu_i \approx 1$, where μ_e and μ_i are the mobility of the electrons and ions, respectively.

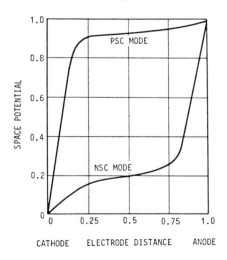

Figure 4.10: Special potential distributions for a glow discharge in a presence of a crossed electro-magnetic field.

The magnetic field prolongs the electron path between electrodes, and thus increases the probability of electron- neutral collisions through which the neutral molecules can be ionized. This enables the discharge of both the PSC and NSC modes to be sustained at a low gas pressure where the mean free path of the electrons is longer than the electrode spacing, i.e., $\lambda > L$ or $p < \lambda_0/L$. Here, λ denotes the electron mean free path at a given pressure p, λ_0 the electron mean free path at 1 Torr, and L the electrode spacing. At such low pressure the discharge cannot be sustained without the magnetic field. Thus, both modes may be used for low pressure sputtering. Kay suggests that at low gas pressure, the PSC mode is much more applicable for sputtering than the NSC mode since the former has a higher sputtering speed than the latter due to the large cathode fall (6). He maintains that in order to have the large cathode fall, the magnetic field should be as small as possible. The optimum magnetic field is then determined by the electron cut off field below which the glow discharge cannot be sustained.

The properties of the NSC mode are somewhat modified at low gas pressure under the influence of a very strong magnetic field. Fig. 4.11 shows the potential distribution across the electrodes (5). Significant voltages at both the cathode fall and the anode fall are seen. In this modified mode, the electron can gain the ionization energy in both falls. The cathode fall results from the loss of electrons due to their recapture at the cathode surface. Since the large cathode fall results in significant levels of sputtering, this mode is useful for many practical sputtering applications (7,8). In this mode the strong magnetic field can increase the ionization efficiency remarkably. Thus the operating pressure can be as low as 10^{-5} Torr or less for an electrode spacing of about 1 cm. This modified mode can be adapted to a practical low pressure sputtering system.

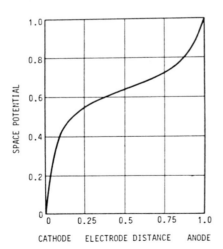

Figure 4.11: A spatial potential distribution in a strong magnetic field.

For argon discharge with cylindrical electrodes where the radii of the cathode and anode are 10 and 30mm, respectively, a transition magnetic field where the PSC mode changes into the NSC mode is roughly estimated from the following relationship $v_i/v_e \simeq 1$, where v_i denotes the radial ion velocity and v_e the radial electron velocity. In low pressure such as 1 mTorr, the ion radial velocity depends on the square root of the discharge voltage since the ion mean free path is longer than the electrode spacing and the ion mass is so great that the magnetic field hardly causes a deflection of its path. Then the ion velocity v_i is given by $V_i \simeq (2eV_s/m_i)^{1/2}$, where e denotes the electronic charge, V_s the discharge voltage, and m_i the mass of the ions. The electron drifts in the radial direction due to collisions with the gas molecules, and V_e is given by $V_e \simeq 16m_eE^2/\pi^2 e\lambda B^3$, where m_e denotes the mass of the electrons, E the radial electric field, and B the strength of a given magnetic field. $V_i/V_e = 1$ becomes $(2eV/m_i)^{1/2}(\pi^2 e\lambda B^3/16m_e E^2) = 1$. Putting $e = 1.6\text{x}10^{-19}\text{C}$, $\lambda \simeq .2\text{m}$ at 1 mTorr, $m_e = 9.1\text{x}10^{-31}\text{kG}$, $V_s \simeq 1500\text{V}$, $m_i = 6.6\text{x}10^{-26}\text{kg(Ar)}$ $(E \simeq 7.5\text{x}10^4\text{V/m})$, we have found that the transition magnetic field is estimated to be about 140G. The modified NSC mode may appear in a much higher magnetic at a magnetic field of more than 1000G, regardless of the electrode size.

4.1.2.4 Plasma in a Glow Discharge: In the positive column of the glow discharge there is a plasma composed of the same number of electrons and ions. The energy of electrons and ions in the plasma is estimated as follows: the electrons (mass m) and ions (mass M) in the plasma, whose initial velocity is zero, are accelerated by the electric field E with an acceleration rate of Ee/m and Ee/M, respectively. The energy given for the electron and ion from the electric field E in time t is expressed by

$$(Eet)^2/2m \quad \text{(electron)}, \tag{4.24}$$

$$(Eet)^2/2M \quad \text{(ion)}.$$

Since m<<M, most of the energy is transferred to the electron from the electric field in the initial stage. In low gas pressure discharge, the collision rate between electrons and gas molecules is not frequent enough for a non-thermal equilibrium to exist between the energy of the electrons and the gas molecules. So, the high energetic particles are mostly composed of electrons while the energy of the gas molecules is around room temperature. We have $T_e >> T_i >> T_g$, where T_e, T_i and T_g are the temperatures of the electron, ion, and gas molecules, respectively. This type of plasma is called "cold plasma".

In a high pressure gas discharge, the collisions between electrons and gas molecules occur frequently. This causes thermal equilibrium between the electrons and the gas molecules. We have $T_e \simeq T_g$. We call this type of plasma as "hot plasma".

In cold plasma, the degree of ionization is below 10^{-4}. The electrons receive energy from the electrical field and collide with the neutral atoms and/or molecules. This results in excitation and/or ionization of the atoms and gas molecules.

In plasma assisted deposition systems, thin films are grown on the negatively biased electrode. The negative bias is induced even in rf discharge, since electron mobility is much higher than that of ions. The negative bias is induced at the rf electrode with the blocking capacitor, as shown in Fig. 4.12.

The induced negative biased V induced is estimated as follows: The positive ions of mass M come from the plasma and traverse the dark spaces without making any collisions, and form a the space charge region near the electrode with a current density j:

$$j = \frac{KV^{3/2}}{M^{1/2}d^2} , \qquad (4.25)$$

where d is the thickness of the space charge zone and K is a constant.

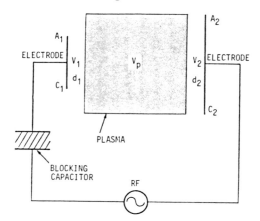

Figure 4.12: Potential induced on electrodes which are immersed in a plasma.

The total current flow is equal at both electrodes, therefore

$$j_1 A_1 = j_2 A_2. \qquad (4.26)$$

Combining this with (4.25),

$$\frac{A_1 V_1^{3/2}}{d_1^2} = \frac{A_2 V_2^{3/2}}{d_2^2} . \qquad (4.27)$$

The capacitance across the dark space is proportional to the electrode area and inversely proportional to the dark space thickness. The rf voltage is capacitively divided between the two sheaths near the electrodes: we have

$$\frac{V_1}{V_2} = \frac{c_2}{c_1} = \frac{A_1}{d_1} \frac{d_2}{A_2} , \qquad (4.28)$$

where c_1 and c_2 are the capacitances across the dark space for each electrode. Substituting this into (4.27),

$$\frac{V_1}{V_2} = \left(\frac{A_2}{A_1} \right)^2 . \qquad (4.29)$$

This suggests that the self-bias voltage ratio V_1/V_2 is proportional to the square of the inverse area ratio A_2/A_1. It is noted that under an assumption that the current density of the positive ions is equal at both electrodes, the V_1/V_2 is proportional to $(A_2/A_1)^4$. However, the relation (4.29) shows the actual power dependence.

4.2 SPUTTERING SYSTEMS

Several sputtering systems for the purpose of thin film deposition. Their designs are shown in Fig. 4.13. Among these sputtering systems the basic model is the dc diode sputtering system. The other sputtering systems are improved systems of the dc diode sputtering.

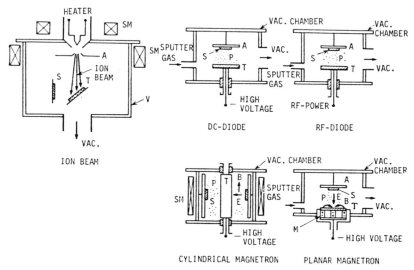

Figure 4.13: Basic configurations of sputtering systems.

4.2.1 DC Diode Sputtering

The dc diode sputtering system is composed of a pair of planar electrodes. One of the electrodes is a cold cathode and the other is an anode. The top plasma-facing surface of the cathode is covered with a target material and the reverse side is water-cooled. The substrates are placed on the anode. When the sputtering chamber is kept in Ar gas at 0.1 Torr and several kilovolts of dc voltage with series resistance of 1 to 10kΩ are applied between the electrodes, the glow discharge is initiated. The Ar ions in the glow discharge are accelerated at the cathode fall and sputter the target resulting in the deposition of thin film on the substrates.

In the dc diode system, sputtered particles collide with gas molecules and then eventually diffuse to the substrate since the gas pressure is so high and the mean free path of the sputtered particles is less than the electrode spacing. The amount of sputtered material deposited on a unit substrate area W is then given by

$$W \simeq k_1 W_0/pd, \qquad (4.30)$$

and the deposition rate R is given by

$$R = W/t, \qquad (4.31)$$

where k_1 is a constant, W_0 is the amount of sputtered particles from the unit cathode area, p is the discharge gas pressure, d is the electrode spacing, W is the density of the sputtered films, and t is the sputter time.

The amount of sputtered particles from the unit cathode area W_0 is given by

$$W_0 = (j_+/e)st(A/N), \qquad (4.32)$$

where j_+ is the ion current density at the cathode, e is the electron charge, S is sputter yield, A is atomic weight of sputtered materials, and N is Avogadro's number. With the assumption that the ion current is nearly equal to discharge current I_s and the sputter yield is proportional to the discharge voltage V_s, the total amounts of sputtered particles become $V_s I_s t/pd$. Thus, the sputtered deposit is proportional to $V_s I_s t$

4.2.2 Rf Diode Sputtering

By simple substitution of an insulator target for the metal target in a dc diode sputtering system, the sputtering glow discharge can not be sustained because of the immediate build-up of a surface charge of positive ions on the front side of the insulator. To sustain the glow discharge with an insulator target, the dc voltage power supply is replaced by an rf power supply. This system is called an rf sputtering system.

Sputtering in the rf-discharge was observed by Robertson and Clapp in 1933 (9). They found that the glass surface of the discharge tube was sputtered during the rf-discharge. In the 1960's, sputtering in the rf-discharge has been used for the deposition of dielectric thin films and a practical rf-sputtering system was developed (10,11). Presently the rf-sputtering system holds an important position in the deposition of thin films. A typical sputtering system is shown in Fig. 4.14.

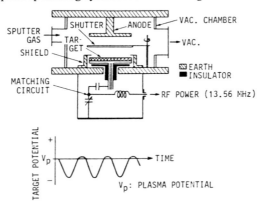

Figure 4.14:   Rf-diode sputtering system.

The rf diode sputtering system requires an impedance-matching network between the power supply and discharge chamber. A typical network for impedance matching is shown in Fig. 4.15 (12).

The impedance of the rf-power supply is almost always 50Ω. The impedance of the glow discharge is of the order of 1 to 10kΩ.

In rf diode sputtering, the target current density i_s is given by

$$i_s \simeq C \, dV/dt \qquad\qquad (4.33)$$

where C is capacitance between discharge plasma and the target, dv/dt denotes the time variations of the target surface potential. This indicates that the increase of the frequency increases the target ion currents. In practical systems, the frequency used is 13.56MHz.

Note that in the rf discharge system the operating pressure is lowered to as low as 1 mTorr, since the rf electrical field in the discharge chamber increases the collision proba- bility between secondary electrons and gas molecules. In the rf sputtering system, a blocking capacitor is connected between the matching network and the target. The target area is much smaller than the grounded anode and the chamber wall. This asymmetric electrode configuration induces negative dc bias on the target, and this causes sputtering in the rf system. The dc bias is on the order of one half of the peak-to-peak voltage of the rf power supply.

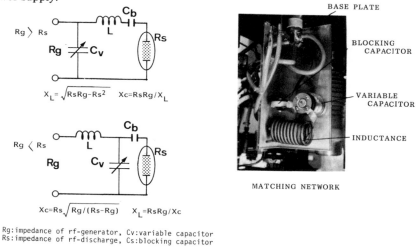

Rg: impedance of rf-generator, Cv: variable capacitor
Rs: impedance of rf-discharge, Cs: blocking capacitor

Figure 4.15: Impedance matching networks for rf-sputtering system.

Figure 4.16 shows the photograph of the sputtering target during the deposition. In the rf-sputtering systems the target and inductance in the matching network are always cooled by water. The electrical resistivity of the cooling water should be high enough as to serve as electrical insulation.

THERMO COUPLES   SUBSTRATE HOLDER

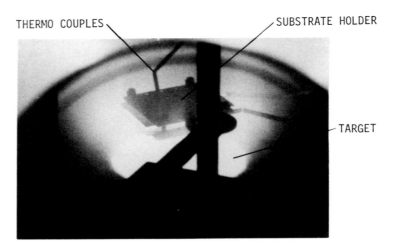

TARGET

Figure 4.16: Photograph showing the rf-sputtering system during a deposition.

4.2.3 Magnetron Sputtering

Low pressure sputtering is one of the most promising techniques for the production of thin film devices. A wide variety of thin films can be made with little film contamination and at a high deposition rate by this technique.

In 1935 Penning first studied low pressure sputtering in which a transverse magnetic field was superposed on a dc glow discharge tube as shown in Fig. 4.17 (13). The experimental system was composed of coaxial cylindrical electrodes with an axial magnetic field, similar to a cold cathode magnetron. He found that superimposition of the magnetic field of 300G lowered the sputtering gas pressure by a factor of ten, and increased the deposition rate of sputtered films. However, this kind of system was not used in practice.

In the early 1960's, magnetron sputtering was reconsidered as an attractive process for thin film deposition by a few workers. Kay studied the glow discharge in the presence of a magnetic field in relation to thin film deposition (14,15). Gill and Kay proposed an inverted magnetron sputtering system and demonstrated that the sputtering gas pressure was as low as 10^{-5} Torr, which was two orders lower than conventional sputtering systems. The strength of the magnetic field was several hundred gauss and the PSC (positive space charge) mode was dominant in the sputtering discharge.

Hayakawa and Wasa also studied this type of discharge in relation to glow discharge mode (16), plasma instability (17), and cathode sputtering (18-20). They invented an original planar magnetron sputtering system with a solenoid coil (21). Figure 4.18 shows the construction of the system (21). It was also found that the presence of a strong magnetic field changed the glow discharge mode; i.e. the NSC (negative space charge) mode appeared and enhanced the cathode sputtering. However in those days, this type of magnetron sputtering system was not widely used in practice.

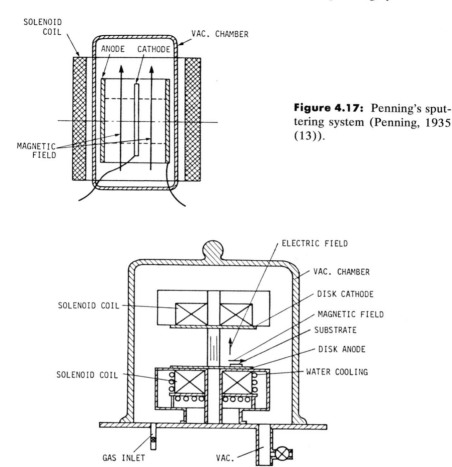

Figure 4.17: Penning's sputtering system (Penning, 1935 (13)).

Figure 4.18: Construction of a planar magnetron sputtering system with solenoid coil (21).

In the 1970's, magnetron sputtering become more widely used for its potential applications for Si integrated circuits, as well as for various other thin film coating processes (22,23). Today, it is in broad scale practical use (24) in industries ranging from microelectronics, to automobiles, to architectural glass to hard coatings.

Two types of magnetron sputtering systems are widely used for thin film deposition. One is a cylindrical type, the other is a planar type. Typical cathode configurations are shown in Fig. 4.19. Within the cathode target, permanent magnets are embedded such that the resultant magnetic field is several hundred gauss. The glow discharge is concentrated in the high magnetic field region, thus a circular cathode glow is observed as shown in Fig. 4.20. The surface of the cathode is nonuniform due to the circular cathode glow. This shortens the actual life of the of the cathode target. Several types of improved magnetron targets have been proposed including a magnetron target with moving magnets or multi-magnets in order to have an uniform erosion area and also to extend the actual life of the target (Fig. 4.21).

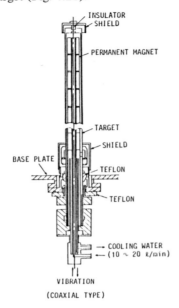

Figure 4.19: Construction of magnetron target.

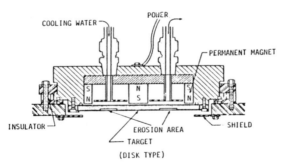

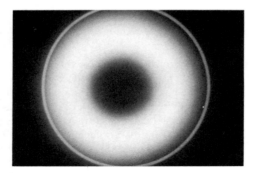

Figure 4.20: Photograph showing a top view of a planar disk target in a magnetron discharge; target 100mm diameter.

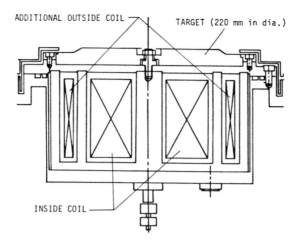

Figure 4.21: Construction of a magnetron target with an additional solenoid coil.

Magnetron sputtering sources have been found to be of limited use when magnetic target materials are used. The magnetic field lines are confined within the target material unless very thin targets are used where magnetic saturation can be achieved. Chopra and Vankar have designed and fabricated a sputtering cathode which offers both of these advantages and, in addition, yields excellent uniformity of the deposited films without employing a cumbersome substrate rotation. This geometry can also be used with magnetic materials with equal ease.

Figure 4.22 shows the construction of the planar magnetron sputtering source designed by Chopra and co-workers (25). It consists of a water-cooled cathode (A) made of copper on which planar targets (B) of any material bounded to a copper backing plate can easily be screw mounted. The cathode is insulated from a water-cooled aluminum shield (C) with a Teflon spacer (D) and is kept in position using a stainless-steel nut (I). The magnets (J) are mounted outside the shield. The whole assembly is affixed to a stainless-steel base plate (G) which is placed in a bell jar in which sputtering is carried out. The discharge voltage is 300 to 800V where the maximum sputtering yield per unit energy input is obtained.

In the magnetron sputtering system the working pressure is 10^{-5} to 10^{-3} Torr, and the sputtered particles traverse the discharge space without collisions. Thus the deposition rate R is simply given by

$$R \simeq kW_0/t, \tag{4.34}$$

where k=1 for the planar system, $k = r_c/r_a$ for the cylindrical system and r_c is the cathode radius and r_a is the anode radius. W_0 is the amount of sputtered particles given by the relationship (4.30) (19).

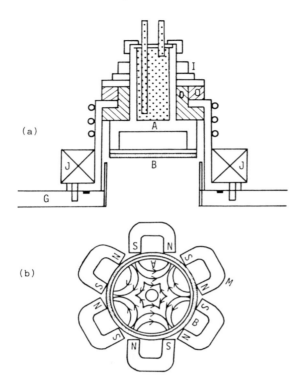

(a)

(b)

Figure 4.22: Cross-sectional view (a) and top view (b) of the planar magnetron sputtering source. (A) cathode, (B) planar target, (C) aluminum shield, (D) teflon spacer, (I) stainless-steel nut, (J) magnets, and (G) stainless-steel baseplate (Rastogi, Wanker and Chopra, 1987 (25)).

Typical experimental results are shown in Figure 4.23.

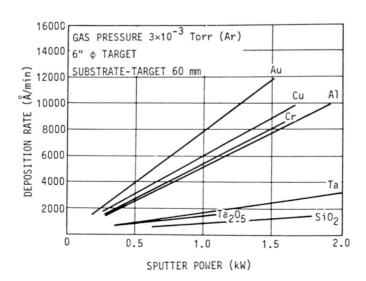

SPUTTER POWER (kW)

Figure 4.23 (a) Deposition rate as a function of sputtering power for a planar magnetron sputtering system, dc-magnetron for a metal target, rf-magnetron for a dielectric target.

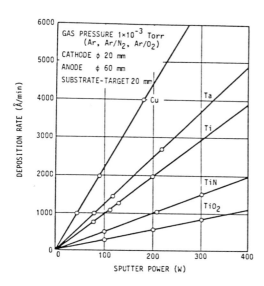

Figure 4.23: (b) Deposition rate as a function of sputtering power for a dc coaxial magnetron.

4.2.4 Ion Beam Sputtering

In the glow discharge system the gas pressure is so high that the sputtered films are irradiated by gas molecules during thin film growth. This causes the inclusion of gas molecules in the sputtered films. In the ion beam sputtering system the incident ions are generated at the ion source. The target is bombarded by the ion beam in a sputtering chamber separated from the ion source.

Pioneering work was done by Chopra and Randlett on the deposition of thin films by ion beam sputtering (26). They constructed the ion beam sputtering system and showed how it could be used for deposition of thin films of metals and insulators. Their sputtering system is shown in Fig. 4.24. The ion source is composed of the von Ardene type of duoplasmatron in which an arc discharge is maintained so as to create the ions for the sputtering. A well defined ion beam with a current of up to $500\mu A$ over an area $\simeq 1 cm^2$ can be extracted into the vacuum chamber at a pressure of 10^{-5} Torr with an accelerating voltage of 1 to 2 kV.

More recently, Kaufman-type broad beam ion sources have been developed for use in thin film applications. This type of source typically has a plasma chamber within the ion source with a hot filament cathode. Ions created in the source can be accelerated by means of multi-aperture grids to form a large area, intense beam.

The ion beam current ranges from 10 mA to several Amperes, depending on the dimensions of the source. The ion energy can be varied from 0.5 to 2.5 kV. The gas pressure of the sputtering chamber is typically in the high 10^{-5} to low 10^{-4} Torr range. This kind of system is widely used for sputter etching of semiconducting devices. Increasing interest has been recently paid to ion beam sputtering not only for semiconducting processes but also for developing exotic materials. Figure 4.25 shows a conventional ion beam sputtering system with Kaufman ion source.

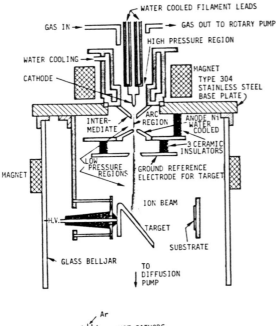

Figure 4.24: Construction of the duoplasmatron argon ion source for sputtering deposition (Chopra and Randlett, 1967 (26)).

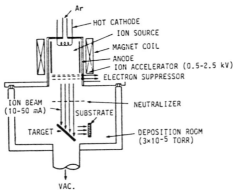

Figure 4.25: Construction of conventional ion beam sputtering system (Veeco).

4.2.5 ECR Plasmas

ECR (electron cyclotron resonance) microwave-based plasmas have been developed primarily for applications with reactive etching or deposition. However, a variation of the source can be configured for sputter deposition or reactive sputter deposition. The ECR discharge is sustained under rf-electric field with static magnetic field. The ECR conditions are given by

$$f = 1/2\pi \frac{eB}{m},$$ (4.35)

where f denotes the frequency of the rf-electric field, B, magnetic field strength, and e and m, electron charge and mass, respectively. For the f = 2.45GHz, the B becomes 874G.

A typical ECR type sputtering system is shown in Fig. 4.26. The system comprises negatively biased ring target settled at the outlet of the discharge chamber. The target bias is 0.4 to 1 keV. Since the system uses a chemically stable cold cathode, reactive gases could also be used for the sputtering. The operating pressure is as low as 10^{-5} Torr so the ions sputter the surface deposit atoms onto samples in a line-of-sight mode. Table 4.4 shows the summary of the operative properties of these sputtering systems (27).

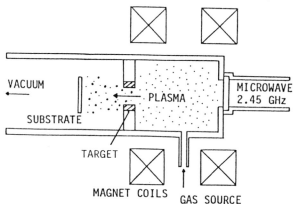

Figure 4.26 ECR plasma sputtering system with ring target.

4.3 PRACTICAL ASPECTS OF SPUTTERING SYSTEMS

When using the sputtering systems several kinds of equipment are prepared. These include: sputtering targets, sputtering gas, substrate holders and monitoring systems.

4.3.1 Targets for Sputtering

The target is generally made of a metal or alloy disk. Hot-pressed sintered disks are used for depositing compound thin films. The diameter is 5 to 8 cm when used for re-search, and 15 to 30 cm for production. A rectangular plate is also used for a production. Typical sputtering targets are listed in Table. 4.5. Figure 4.27 shows a photograph of the targets.

Sputtering is an inefficient process, and most of the power input to the system appears ultimately as target heating. Thus these targets are usually mounted on a water-cooled backing plate. The target is fixed on the backing plate by mounting clips or mechanical support. It is surrounded by a dark space shield, known as a ground shield, so as to be sputtered target material only. The construction of the target with the shield is shown in Fig. 4.28. The spacing between the target and the ground shield must be less than the thickness of the dark space λ_0/p. Taking $\lambda_0 = 0.05$cm, p = 10 mTorr, λ_0/p becomes 5 cm.

The condition of the target strongly affects the properties of the resultant thin films. Note that the resistivity of the cooling water should be high enough so as to keep an electrical insulation between the target and the grounded chamber.

Operating Properties		DC, RF-diode Sputter	Planar Magnetron	Ion Beam Sputter	ECR Sputter
operating pressure		$0.1 \sim 100\,\text{Pa}$	$0.01 \sim 10\,\text{Pa}$	$0.001 \sim 0.1\,\text{Pa}$	$0.001 \sim 0.1\,\text{Pa}$
ionization degree		$10^{-5} \sim 10^{-4}$	$10^{-5} \sim 10^{-2}$	--	$10^{-3} \sim 10^{-1}$
electron temperature		$10^{4} \sim 10^{5}\,\text{K}$	$10^{4} \sim 10^{5}\,\text{K}$	--	$5 \times 10^{4} \sim 10^{6}\,\text{K}$
ion temperature		$\sim 10^{3}\,\text{K}$	$\sim 10^{3}\,\text{K}$	--	$10^{3} \sim 10^{4}\,\text{K}$
gas temperature		$\sim 10^{3}\,\text{K}$	$\sim 10^{3}\,\text{K}$	--	$\sim 10^{3}\,\text{K}$
plasma density		$10^{8} \sim 10^{10}\,\text{cm}^{-3}$	$10^{9} \sim 10^{12}\,\text{cm}^{-3}$	--	$10^{9} \sim 10^{12}\,\text{cm}^{-3}$
particle energy	adatoms	$1 \sim 10\,\text{eV}$	$1 \sim 10\,\text{eV}$	$1 \sim 10\,\text{eV}$	$1 \sim 10\,\text{eV}$
	secondary electron bombardment	large	small	--	small
ionization degree of adatoms		$10^{-3} \sim 10^{-2}$	$10^{-3} \sim 10^{-1}$	$10^{-3} \sim 10^{-2}$	$10^{-2} \sim 10^{-1}$

Table 4.4: Operating properties of sputtering systems.

TARGET	PURITY (%)			TARGET	PURITY (%)		
Ag		99.99		Ni	99.9		
Al		99.99	99.999	Pd		99.99	99.999
Au		99.99		Pt		99.99	
Bi	99.9	99.99	99.999	Sb			99.999
C			99.999	Si			99.999
Co	99.9			Sn			99.999
Cr	99.9	99.99		Ta		99.99	
Cu		99.99		Ti	99.9		
Fe	99.9			V	99.9		
Ge			99.999	W		99.99	
Hf	99.9			Y	99.9		
In		99.99		Zn			99.999
Mo		99.99		Zr	99.9		
Nb		99.99					

TARGET	PURITY (%)				TARGET	PURITY (%)			
AlN	99				Mo-Si$_2$	99			
Al$_2$O$_3$		99.9	99.99		NbN	99			
B$_4$C	99				Nb$_2$O$_5$		99.9	99.99	
BN	99				PbO				99.999
BaTiO$_3$	99				PbS		99.9		
Bi$_2$O$_3$		99.9			PbTiO$_3$		99.9		
Bi$_2$Te$_3$				99.999	SiC	99			
CdS				99.999	Si$_3$N$_4$	99	99.9		
CdTe				99.999	SiO$_2$			99.99	
CrSi$_2$	99				SnO$_2$		99.9	99.99	
Cu$_2$S			99.99		TaS$_2$		99.9		
Fe$_2$O$_3$		99.9	99.99	99.999	TiC	99			
Fe$_3$O$_4$		99.9			TiN	99			
HfO$_2$		99.9			WO$_3$			99.99	
In$_2$O$_3$			99.99		Y$_2$O$_3$			99.99	
ITO	(In$_2$O$_3$)$_{0.8}$	(SnO$_2$)$_{0.2}$			ZnO				99.999
LiNbO$_3$		99.9	99.99		ZnS			99.99	99.999
LiTaO$_3$		99.9	99.99		ZnSe				99.999
MgO		99.9	99.99		YBaCuO				
					LaSrCuO				

Table 4.5: Sputtering targets and purities.

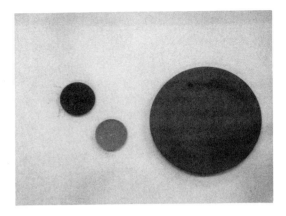

Figure 4.27: Photographs of some sputtering targets.

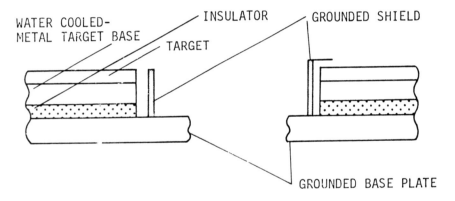

Figure 4.28 Construction of target with ground shield.

4.3.1.1 Compound Targets: Thin films of alloys can be sputtered from a composite target as shown in Fig. 4.29. The composition of the thin films is controlled by the area ratio of each element.   Taking the sputtering yield and area ratio of each element, (s_1, a_1), (s_2, a_2), (s_3, a_3), ... (s_n, a_n), partial composition of each element becomes $s_1 a_1 (A_1/N)$, $s_2 a_2 (A_2/N)$, $s_3 A_3 (A_3/N)$, ...$s_n A_n (A_n/N)$ where A_1, A_2, A_3... denote the atomic weights of each element, and N is Avogadro's number.

When the number of partitions is small the composition of the sputtered films will distribute nonuniformly over the substrate. A rotating substrate holder is often used to obtain uniform composition over the entire substrate.

Hanak suggested that a binary compound target achieved deposition of thin films with various composition ratios (28). Figure 4.30 shows the binary compound target. It suggests thin films of different composition are obtained at different substrate positions.

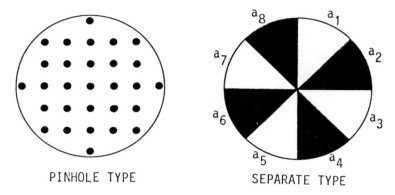

PINHOLE TYPE   SEPARATE TYPE

Figure 4.29 Construction of composite target for the deposition of alloy, compound thin films.

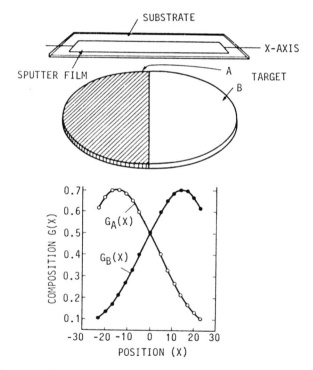

Figure 4.30: A composite target for depositing a binary alloy (Hanak, 1971 (28)).

4.3.1.2 Powder Targets: Thin films of compounds including metal oxides, nitrides, carbides can be deposited by direct sputtering from the sintered powder of these compounds. The sintered powder is filled in a stainless dish which is mounted on the backing plate. A wide variety of compound thin films can be sputtered from a powder target. The construction of the powder target with the stainless dish is shown in Fig. 4.31.

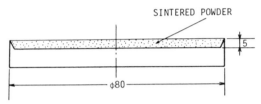

Figure 4.31: Powder target and a construction of the stainless dish for the powder target.

4.3.1.3. Auxiliary cathode: Small amounts of foreign metals can be mixed to thin films by co-sputtering of an auxiliary cathode made of foreign metals. Variations of the sputtering current to the auxiliary cathode change the amounts of foreign metals in the resultant films (29).

4.3.2 Sputtering Gas

For the deposition of metals, pure Ar gas (purity, 99.99 to 99.9999%) is introduced through a variable leak valve. An automatic gas flow controller is also usually used for the sputtering system. A typical gas flow system is shown in Fig. 4.32. The vacuum system should be water vapor free. The liquid nitrogen trap should be used for the oil diffusion system.

When we use a reactive gas such as oxygen, thin films of metal oxides can be deposited from a metal target, and is known as reactive sputtering. In reactive sputtering, the reaction will be taken place both at the cathode surface and the substrate during deposition. An impingement of the reactive gas of the cathode surface forms the compounds such as metal oxides, and the resultant compounds will be sputtered. This leads to deposition of compounds.

The optimum concentration of the reactive gas is determined by

$$kN_g/N_c > 1 \qquad \text{reaction at cathode,} \qquad (4.36)$$

$$kN_g/N_s > 1 \qquad \text{reaction at substrate,}$$

where N_g denotes the number of reactive gas molecules which strike a unit area of the cathode surface or the substrate per unit time, N_c is the number of sputtered atoms from the unit area of cathode per unit time, and N_s is the number of deposited atoms per unit area of substrate per unit time.

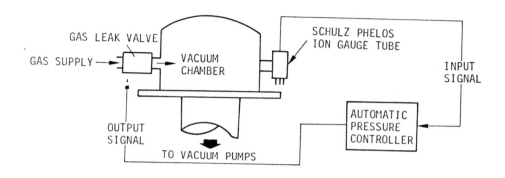

Figure 4.32: Typical gas flow system for the sputtering system.

When we take the case of sputtering of Ti cathode in the reactive gas O_2, N_g is $3.5 \times 10^{20} P_{O_2}$ molecules/sec/cm at the partial oxygen pressure of P_{O_2} in Torr and $k \approx 1$. Putting these relationships into (4.36), a minimum oxygen pressure for the reactive sputtering is given by the following relationship:

$$P_c \approx N_c/(3.5 \times 10^{20}),$$

(4.37)

$$P_s \approx N_s/(3.5 \times 10^{20}),$$

where P_c is the minimum oxygen pressure for the cathode surface reaction, and P_s the minimum oxygen pressure for the substrate surface reaction. Taking $N_c \approx 3 \times 10^{16}$ atoms/sec/cm², $N_s \approx 1 \times 10^{16}$ atoms/sec/cm² at a deposition speed of 120 Å/min, we have $P_c = 8.5 \times 10^{-5}$ Torr and $P_s \approx 3 \times 10^{-5}$ Torr (30).

Since a TiO_2 film is produced when the cathode surface layer is fully oxidized to TiO_2 by oxygen absorption, stoichiometric TiO_2 films can be produced when $P_{O_2} > 3 \times 10^{-5}$ Torr. On the other hand, since metallic Ti films may be produced when neither cathode surface nor the sputtered film are fully oxidized, the metallic Ti films can be produced when $P << 3 \times 10^{-5}$ Torr. Generally we use the mixed gas of Ar and reactive gas for the reactive sputtering. The concentration of the reactive gas is generally 5 to 50%. Table 4.6 shows the typical discharge gas for the sputtering deposition. These sputtering gases are uniformly fed to the system through a stainless steel pipe or teflon pipe with fitting components as shown in Fig. 4.33. The presence of nonuniformity in reactive gas density in the system results in nonuniformity in the chemical composition of resultant films.

Gas	Class	Purity %	P kg/cm²	N₂	Ar	Impurities, (ppm) O₂	CH₄	CO	CO₂	NOₓ	Dew Point
N₂	pure S	99.9998	150			<.5	<.1	<.5	<.1	<.01	<-70C
	pure A	>99.9995	150			<.5	<.5				<-70C
	pure B	>99.9995	150			<1					<-70C
	stand.	99.999	150			<5					<-65C
Ar	pure A	<99.9995	150	<3		<.2	<1		<1		<-70C
	pure B	<99.999	150			<.2	<1		<1		<-70C
	stand.	99.998	150	<10		<2					<-65C
O₂	pure A	>99.99	50	<20	<50		<30	<10	<10	<10	<-70C
	pure B	>99.9	150	<1k	<1k		<30				<-60C
	stand.	99.6	150	<400	<400						<-65C
H₂	pure	>99.99999		150		<.05					<-70C
	1st cls	>99.99	150								
NH₃	pure	99.995		<10		<5	<5				H₂O < 10ppm
	1st cls	>99.99	2-10								
CH₄	pure	<99.95	50								
	1st cls	>99	100								
H₂S		>99	7-30								

Nippon Sanso Catalog G 1 (80-6) 3000 T, G 2 (81-4) 4000M

Table 4.6: Sputtering gasses used and typical impurity levels.

Figure 4.33: Fitting components for a sputtering gas flow system.

4.3.3 Thickness Distribution

Strictly speaking the thickness distribution of sputtered films is governed by several factors including the angular distribution of sputtered particles, collisions between sputtered particles and gas molecules, and the construction of the target shield. However the thickness distribution is estimated by the simple assumption that the angular distribution obeys the cosine law similar to the vacuum evaporation and the collisions are neglected for the sputtered particles.

In the case of the diode sputtering system as shown in Fig. 4.34, the thickness distribution is estimated for a disk cathode target (first equation) and a ring cathode target (second equation):

$$d/d_o = \frac{1 + (S/h)^2}{2(S/h)^2} \left[1 - \frac{1 + (L/h)^2 - (S/h)^2}{\sqrt{[1 - (L/h)^2 + (S/h)^2]^2 + 4(L/h)^2}}\right], \qquad (4.38)$$

$$d/d_o = [1 + (S/h)^2]^2 \frac{1 + (L/h)^2 + (S/h)^2}{[[1 - (L/h)^2 + (S/h)^2]^2 + 4(L/h)^2]^{3/2}},$$

where d_o is the center thickness, d the thickness at a center distance L, s the radius of the disk and ring cathode, and h the evaporation distance (31). The thickness distribution for conventional diode sputtering is estimated by the disk target system. For the planar magnetron, the ring target system is used for estimating the thickness distribution. Figures 4.35 and 4.36 show typical results for a disk target and a ring target, respectively.

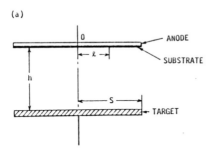

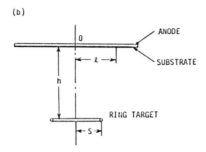

Figure 4.34: Construction of diode sputtering system. (a) planar electrode with disk target, (b) ring target.

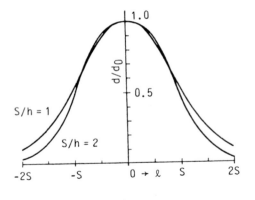

Figure 4.35: Thickness distribution for the disk target.

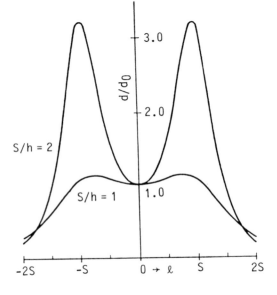

Figure 4.36: Thickness distribution for the ring target.

4.3.4 Substrate Temperature

The temperature of the substrate surface is an important and yet difficult parameter to control. In conventional sputtering systems, the substrate is mounted on a temperature controlled substrate holder. However, the surface of the substrate is heated by the radiant heat of the target. Moreover, bombardment by high energy secondary electrons also heat the substrate. In the rf diode system the temperature of the substrate rises up to 700°C without additional substrate heating. In order to reduce the effect of the radiant heat, the surface of the target must be cooled. Bombardment by the secondary electrons is avoided by negatively biasing the substrate.

It is noted the temperature rise of the substrate depends on the type of the sputtering system. The rf diode system shows the highest temperature rise and the magnetron system shows the lowest temperature rise. The selection of the type of the sputtering system is important. The typical temperature rise for rf-sputtering systems is shown in Fig. 4.37.

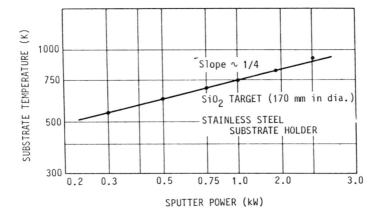

Figure 4.37: Temperature rise of substrates during deposition for rf-sputtering system.

4.3.5 Monitoring

The monitoring of sputtering conditions during deposition is important in order to control properties of the resultant films.   Parameters to be monitored are as follows:

1.   discharge voltage, current, power,

2.   residual gas,

3.   partial pressure of the sputter gas,

4.   substrate temperature,

5.   thickness of the sputtered films.

The quadrupole mass analyzer or optical spectrometer are widely used for monitoring the composition of the sputter gas. The thickness of transparent films is monitored by the laser interference method. Langmuir probes are used for monitoring discharge or plasma parameters including spatial distribution of potential, electron density and electron temperature.

4.3.5.1 Gas Composition: Figure 4.38 shows the construction of the quadrupole mass analyzer. The quadrupole mass analyzer is composed of an ion source chamber, a focusing electrode, quadrupole electrodes and an ion collector. The electrodes are about 30 cm long and the pairs of opposing rods are connected to a dc and rf voltage supply. When the sputtering gas is introduced into the ion source chamber, the sputtering gas is ionized and the resulting ions are accelerated along the axis between rods. Since, the probability of the number of ions which pass through the quadrupoles and reach the ion collector is governed by the mass number of the sputtering gas and the value of the rf voltage, the mass number is determined by the sweeping of the rf voltage. The photograph of the quadrupole mass analyzer is shown in Fig. 4.39.

The other popular monitoring technique of the sputtering gas is the optical spectrometric method. The emission intensity of a gas species is characteristic of the ionized species in the sputtering gas. The spatial distribution of the ionized species can be estimated by introduction of the optical fiber. Information about the ionized species informs us of the intensity of sputtered atoms. This enables one to monitor the sputtering rate. Recent works suggest that optimum conditions for reactive sputtering can be maintained by the optical spectrometric method. For example, the transparent conductive films, ITO, can easily be produced by sputtering from In-Sn target in an oxidizing atmosphere by monitoring the optical spectrum of In ions. The optical spectrometric method is also used for monitoring the end point of plasma etching.

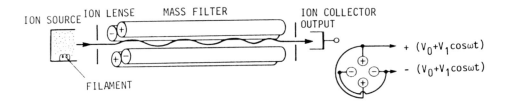

Figure 4.38: Construction of quadrupole mass analyzer.

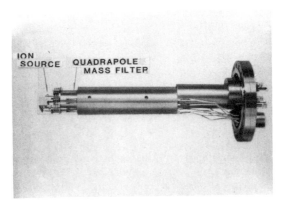

Figure 4.39: Analysis tube of quadrupole mass analyzer.

4.3.5.2. Sputtering Discharge In an rf or dc sputtering discharge the power density of the target is in the order of 1 to 5W/cm^2 At the sputtering voltage of 2,000V, the sputtering current is 0.5 to 2.5mA/cm^2.

At the disk target of 20 cm in radii, the sputtering power is 400 to 2,000W, the sputtering current 200 to 1,000 mA, and the impedance of the sputtering discharge 2 to 10kΩ.

These discharge parameters can be easily measured by a conventional high impedance volt meter and a low impedance current meter for dc sputtering. However, the discharge parameters cannot be exactly measured for rf sputtering. In the conventional rf-sputtering system, the rf power meter is inserted between the matching circuit and the rf power supply. The power loss at the target will be included in the measured values of the rf power meter.

The current and voltage at the target is measured by the thermocouple and the capacitive voltmeter respectively. The exact measurements of these discharge parameters are difficult.

4.3.5.3 Plasma Parameters: Plasma parameters are estimated by the Langmuir probe as shown in Fig. 4.40 (32). The Langmuir probe consists of Mo or W electrodes inserted in the plasma. The plasma parameters are estimated by the current voltage curve of these electrodes. The typical current voltage characteristics are shown in Fig. 4.41.

When the electrode is negatively biased against the plasma potential (a negative probe), a positive space charge is accumulated around the probe and an ion sheath appears. The electric field in the ion sheath repels the electrons from the plasma. For the negative probe, the probe current I_p consists of the ion current I_+ from the plasma. The I_p will, if the Maxwellian rule holds, be given by

$$I^p \simeq I_+ = AeN_+\sqrt{\frac{kT_+}{2\pi M}} \; , \tag{4.39}$$

where A is the probe area, e the electron charge, N_+ the ion density, T_+ the ion temperature, M the mass of ion, and k the Boltzmann constant. Thus the ion density in the plasma will be estimated from the I_p at negative probe voltages (region I).

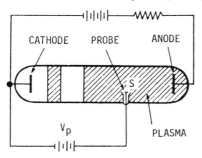

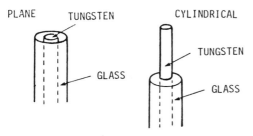

Figure 4.40: Construction of Langmuir probe.

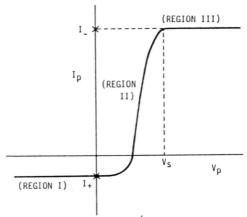

Figure 4.41: Current-voltage characteristics of the Langmuir probe.

When the probe potential is slightly negative to the plasma potential V_s, high energy electrons will flow into the probe against the retarding bias in the ion sheath. The probe current will be given by

$$I_p = I_+ + I_-, \tag{4.40}$$

where I_- denotes the electron current (region II). The I_- is expressed by

$$I_- = AeN_-\sqrt{\frac{kT_-}{2\pi m}} \; exp[- \frac{e(V_p - V_s)}{kT_-}], \tag{4.41}$$

where N_- is the electron density, T_- the electron temperature, m the electron mass, and V_p the probe potential.

Eqn. (4.41) suggests that when $V_p = V_s$, the I_p becomes the electron current I_-. We have

$$I_p \simeq I_- = AeN_-\sqrt{\frac{kT_-}{2\pi m}} \;, \tag{4.42}$$

when $V_p > V_s$ the probe current shows a constant value of I_- (region III).

The electron density will be estimated from the I_p at region III. The space potential V_s will be determined by the transition point from region II to region III.

It is considered that the probe characteristics strongly depend on the construction and surface properties of the metal electrode. The presence of the magnetic field significantly affects the probe characteristics (33). Also note that the plasma should not be disturbed by introduction of the probes. The collisions in the probe sheath are negligible. In other words, the following conditions should be kept for probe measurements.

● The size of probe should be smaller than the mean free-path of the ions and electrons.

● The thickness of the probe sheath should be smaller than their mean free path.

Under these conditions the probe current I_p is expressed by the relationship

$$I_p = AeN \int_{0,u}^{\infty} uf(u)du,$$ (4.43)

where N is the density of the charged particles, u the velocity of the charged particles perpendicular to the probe surface, f(u) their velocity distribution. The lower limit of u is to be taken 0 for the accelerating field (V>0) and $u = (2e(V_s - V_p)/m)^{1/2}$ for the retarding field (V<0). In (4.39) to (4.42) the velocity distribution is considered to be the Maxwellian rule. (4.41) suggests that

$$\frac{d(\log I_-)}{dV_p} = \frac{-e}{kT_-}$$ (4.44)

The $\log I_-$ vs V_p plots show the linear properties and their slope will give the electron temperature T_-.

When we use the Langmuir probe several discrepancies to the probe theory appear. For instance the probe currents in regions I and III will not show saturation. This will result from the variation of the effective probe area due to the change of the ion sheath by the probe voltage.

The presence of the magnetic field will alter the probe characteristics. In a magnetic field in the order of 100 to 500 gauss its effects on ion currents will be negligible. The probe characteristics will affect the electron currents since the electrons will be fixed to the magnetic line of force. In order to avoid the effects of the magnetic field the probe surface should be perpendicular to the magnetic line of force. When the probe surface is parallel to the magnetic line of force, the electron current to the probe is strongly reduced.

Also note that when the magnetic field is superposed onto the discharge the charged particles show the ExB drift motion. The probe current will increase due to the drift motion. The probe current under the drift motion I'_p is given by

$$I_p' = I_p(1 + \alpha^2/2 - \alpha^4/16),$$ (4.45)

where α denotes the ratio of the drift velocity to the thermal velocity.

4.3.5.4 Thickness Monitor: Several methods are proposed for the thickness monitoring during vacuum deposition. Crystal oscillators, resistance monitors, capacitance monitors and optical monitors.

Among these methods the optical interferometric method is one of the most useful methods for thickness monitoring of sputtered films. Typical examples of this method are shown in Fig. 4.42. He-Ne laser light is introduced into the sputtering chamber and the surface of the substrate is irradiated by the laser light during sputtering. When optically transparent films are deposited, the reflected light from the surface of the substrate and the sputtered film will show interference with each other. The thickness is monitored by the periodic properties of the reflected light intensity.

Semitransparent films having a narrow band gap will be monitored by infrared laser. This kind of monitoring method will give the information about the surface roughness during the deposition since the reflected light intensity will decrease for the rough surface of the sputtered films.

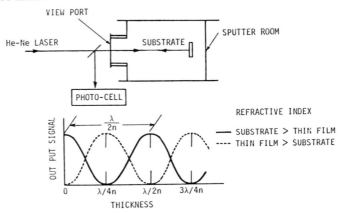

Figure 4.42 Optical interferometric method for monitoring the thickness during the deposition.

4.4 REFERENCES

1. Wasa, K., and Hayakawa, S., Jpn. J. Electrical Engineering, 85: 130 (1965).

2. S.C. Brown, Basic Data of Plasma Physics p. 258 Cambridge, MA, MIT Press (1959).

3. von Engel, A., and Steenbeck, M., Elektrische Gasentladunger Vol II, p.68, Berlin, J. Springer (1932). (J. Springer, Berlin, 1932) Vol.II, p.68.

4. Francis, G., Ionization Phenomena in Gases p. 124, Butterworth Publications (1960).

5. Wasa, K., Hayakawa, S., Proc. IEEE, 55: 2179 (1967).

6. Kay, E., J. Appl. Phys., 34: 760 (1962).

7. Wasa, K., and Hayakawa, S., U. S. Patent 3,528,902 (Sept.1970) assigned to Matsushita Electric Ind. Corp.

8. Wasa, K., and Hayakawa, S., Jpn. Patent 558,099 (1966), assigned to Matsushita Electric. Ind. Corp.

9. Robertson, J.K., and Clapp, C.W., Nature 132: 479 (1933).

10. Anderson, G.S., Mayer, W.N., Wehner, G.K., J.Appl.Phys., 33: 2291 (1962).

11. Davidse, P.D., Maissel, L.I., J. Appl. Phys., 37: 754 (1966).

12. Maissel, L.I., and Glang, R., (eds.) Handbook of Thin Film Technology 4-35, New York, McGraw-Hill (1970).

13. F.M. Penning, U.S. Patent 2,146,025 (Feb. 1935).

14. Gill, W.D., and Kay, E., Rev. Sci. Instrum., 36: 277 (1965).

15. E. Kay, U.S. Patent 309,159 (Sept. 1963), assigned to IBM Corp.

16. Hayakawa, S., and Wasa, K., J. Phys. Soc. Jpn., 20: 1692 (1965).

17. Wasa, K., and Hayakma, S., J. Phys. Soc. Jpn., 20: 1219 (1965).

18. Wasa, K., and Hayakawa, S., IEEE Trans. Parts Materials Packaging, PMP-3: 71 (1967).

19. Wasa, K., and Hayakawa, S., Rev. Sci. Instrum., 40: 693 (1969).

20. Wasa, K., and Hayakawa, S., Thin Solid Films, 52: 31 (1978).

21. K.Wasa and S.Hayakawa, Jpn. Patent 642,012 (1967) assigned to Matsushita Electric Ind. Corp..

22. P.J. Clarke, U. S. Patent 3,616,450 (Oct. 1971).

23. Chapin, J.S., Res./Dev., 25: 37 (1974).

24. Hoffman, V., Solid State Technol. 57, (December 1976).

25. Rastogi, R.S., Vankar, V.D., and Chopra, K.L., Rev.Sci. Instrum. 58: 1505 (1987).

26. Matsuoka, M., and Ono, K., Ohyobutsuri, 57: 1301 (1988).

27. Chopra, K.L. and Randlett, M.R., Rev. Sci. Instr., 38: 1147 (1967).

28. Hanak, J.J., J. Vac. Sci. Technol., 8: 172 (1971).

29. Hada, T., Wasa, K., and Hayakawa, S., Thin Solid Films, 7: 135 (1971).

30. Wasa, K., and Hayakawa, S., Microelectron, Reliab., 6: 213 (1967).

31. Maissel, L.I., and Glang, R., (eds.) Handbook of Thin Film Technology 1-57, New York, McGraw Hill (1970).

32. Langmuir, I., and Blodgett, K.B., Phys. Rev., 22: 347 (1923). ibid., 24: 49 (1924).

33. Hayakawa, S., and Wasa, K., National Tech. Rept., Vol.8, No. 5: 419 (1962).

5

DEPOSITION OF COMPOUND THIN FILMS

Many of the technological advances in the last 20 years in LSI through ULSI integrated circuits can be traced to advances in thin film processing techniques. These advances have allowed the development of many kinds of thin film electronic devices including thin film transistors, surface acoustic devices, high precision resistors, solar-cells, magnetic tape and sensors.

Today, most of these thin films are widely used not only for electronic devices but also for optical coatings, decorations and precision machines parts. Table 5.1 shows typical thin film materials used for these applications. Compound thin films are also important for these practical applications as seen in Table 5.1.

Application		Materials
Electronics	electrodes, interconnections	Au, Al, Cu, Cr, Ti, Pt, Mo, W, Al/Si, Pt/Si, Mo/Si.
	resistor	Cr, Ta, Re, TaN, TiN, NiCr, SiCr, TiCr, SnO_2, In_2O_3.
	dielectrics	AlN, BN, Si_3N_4, Al_2O_3, BeO, SiO, SiO_2, TiO_2, Ta_2O_5, HfO_2, PbO, MgO, Nb_2O_5, Y_2O_3, ZrO_2, $BaTiO_3$, $LiNbO_3$, $PbTiO_3$, PLZT, ZnS.
	insulators	Si_3N_4, Al_2O_3, SiO, SiO_2, TiO_2, Ta_2O_5.
	magnetics	Fe, Co, Ni, Ni-Fe, Te-Fe, GdCo.
	superconductors	Nb, NbN, Nb_3Sn, Nb_3Ge, Nb_3Si, La-Sr-Cu-O, Y-Ba-Cu-O, Bi-Sr-Ca-Cu-O
	semiconductors	Ge, Si, Se, Te, SiC, ZnO, ZnSe, CdSe, CdTe, CdS, PbS, PbO_2, GaAs, GaP, GaN, Mn/Co/Ni/O.
	passivasions	Si_3N_4, SiO, SiO_2,
Optics	coating	SiO_2, TiO_2, SnO_2, In_2O_3.
Instruments, miscellany	hardning, decoration	Cr, TiN, TiC, SiC, WC. Al, Zn, Cd, Cr, Ti, Ta, W, TiN, TiC, SiC. Ag, Au, Al, TiC.

Table 5.1: Thin film materials and applications.

Several methods have been proposed for making compound thin films and the appropriate deposition processes are listed in Table 5.2 (1-86). It is common for multiple authors to report different deposition conditions even for the same deposition method. This is true especially for the deposition of thin compound films, since their controlled deposition process is not well understood.

As seen in Table 5.2 sputtering is the most common process for the deposition of compound thin films. A wide variety of compound films can be deposited by direct sputtering from a compound target or reactive sputtering from a metal target in the presence of reactive gas.

Recent progress in sputtering enables us to make thin films of new ceramics of complex compounds such as $PbTiO_3$ PLZT [(Pb, La)(Zr, Ti)O_3] and high temperature oxide superconductors. Semiconducting thin films of II-VI and III-V groups such as ZnSe and GaAs can also be deposited by sputtering. Alloy thin films of silicides, such as Mo-Si deposited by magnetron sputtering, are used for making the Schottky barrier of MOS devices.

Thin films of high temperature superconductors composed of layered perovskites can successfully be deposited by the sputtering process. Thin films of these newly found materials are important not only for their applications but also for the understanding of the high T_c superconductivity phenomena.

Close control of the thin film growth process is necessary for both scientific research and in applications of thin films. Several attempts have been made to provide a more controlled deposition environment for the thin film growth process. Irradiation with charged particles during film growth is one of the most promising methods to achieve close control of the film growth process. Diamond crystallites can be formed by ion beam sputtering from a graphite target under irradiation of a proton beam onto the film growth surface (87). The irradiation may cause (1) heating up of the surface of the thin films and/or substrate, (2) rearrangements of adatoms, (3) recrystallization, (4) a change in the level of defects or vacancies, and (5) chemical reaction between irradiated particles and adatoms. The effects of irradiation are estimated by calculating collisions between the irradiated particles and adatoms using the 3-dimension Monte Carlo cascade code for sputtering (TRIMSP) (88,89). Several examples of the deposition of thin compound films will be described in following sections.

5.1 OXIDES

5.1.1 ZnO Thin Films

ZnO single crystals show a wurzite hexagonal structure of a wurzite type as shown in Fig. 5.1. These ZnO crystals are known as piezoelectric materials with a large electromechanical coupling factor and a low dielectric constant (90). Typical physical properties of ZnO are listed in Table 5.3.

*) a (amorphous), PC (polycrystal), SC (single crystal)
**) VE (vac. evaporation), IP (ion plating), ARE (activated reactive evaporation), SP (sputter), MSP (magnetron sputter), IBS (ion beam sputter), LA (laser ablation), P-CVD (plasma-CVD)

(1)

Materials	Structure *)	Dep. method **)	Substrate	Dep. conditions Sub. temp. (°C)	Dep. rate (µm/hr)	Miscellanea	Film properties	Ref.
CdS	PC	VE	fuzed quartz	~200	~2		hex. c-axis orientation SAW velocity ~1700 m/s	1)
ZnS	SC	VE	Si (100)	275		ZnS powder alumina crucible	$r_{41}=6.73\times10^{-13}$ m/V (6328 Å) cub.	2)
	SC	RF-SP	NaCl (100)	20	18~54	ZnS target Ar	cub.	
ZnSe	SC	RF-SP	NaCl (100)	290	0.35	ZnSe target Ar	cub.	3)
Al$_2$O$_3$	a	RF-MSP	Si	160~300	0.9~2.1	Al$_2$O$_3$ target Ar/O$_2$=1	$\varepsilon^*\approx9.96$ $n_0=1.61~1.66$ dielectric strength 4×10^6 V/cm	4)
SiO$_2$	a	CVD	fuzed quartz	900~1100	6~150	reaction gas SiCl$_4$+O$_2$	optical wave guide loss 4.5 dB/cm (1.15 µm) 6.4 dB/cm (6328 Å)	5)
	a	RF-MSP	glass	<130	1.2	SiO$_2$ target Ar	sputter gas pressure 1×10^{-3} Torr RF power 4 kW	6)

(2)

Materials	Structure *)	Dep. method **)	Substrate	Dep. conditions Sub. temp. (°C)	Dep. rate (µm/hr)	Miscellanea	Film properties	Ref.
TiO$_2$	a	DC-MSP	glass	RT~200	1.4	Ti target Ar/O$_2$=85/15	$n_0\approx2.5$ (0.5 µm)	6)
	PC	RF-SP	fuzed quartz	350	0.54	ZnO target Ar/O$_2$=8/2	c-axis orientation	7)
	PC	RF-SP	glass	100~200	0.3~0.7	ZnO spherical target Ar/O$_2$=1	c-axis orientation $\gamma<7.5~9.5°$, $\phi<3$	8)
ZnO	PC	RF-MSP	fuzed quartz	250~320	2~3	ZnO target Ar/O$_2$=1	c-axis orientation $\sigma<1°$	9)
	SC	RF-SP	c, R-sapphire	600	0.2	ZnO target Ar/O$_2$=1	$\rho=2.4\times10^3$ Ωcm (c-sapphire) $\rho\approx70$Ωcm (R-sapphire) $\mu_H\approx2.6~28$cm^2/V.s	10)
	SC	RF-MSP	R-sapphire	400	0.11~0.25	ZnO(Li$_2$CO$_3$) target Ar/O$_2$=1	SAW verocity ~5160 m/s $k^2=3.5\%$	11)
Bi$_{12}$GeO$_{20}$ (BGO)	PC	RF-SP	glass	100~350	0.2~0.6	BGO target	$n_0=2.6$	12)
Bi$_4$Ti$_3$O$_{12}$	SC	RF-SP	Pt (001)	700		Bi$_4$Ti$_3$O$_{12}$ target (Bi rich)	$\varepsilon^*\approx120$, Ps=48 µC/cm^2 ferroelectrics	13) 14)

Table 5.2: Deposition methods for compound thin films.

(3)

Materials	Structure *)	Dep. method **)	Substrate	Dep. conditions			Film properties	Ref.
				Sub. temp. (°C)	Dep. rate (µm/hr)	Miscellanea		
$Bi_{12}TiO_{20}$	SC	RF-SP	BGO	425	0.5		Optical wave guide loss 15 dB/cm (6328 Å)	15)
$Bi_{12}PbO_{19}$	PC	RF-SP	glass	100~600	0.6	$Bi_{12}PbO_{19}$ target	piezoelectrics $k_t \approx 0.22$ (470 MHz)	16)
Bi_2WO_6	PC	RF-SP	glass	sputter RT anneal 200	0.4	Bi_2WO_6 target	n=2.5 ferroelectrics	17)
$(In_2O_3)_{0.8}(SnO_2)_{0.2}$ (ITO)	PC	RF-MSP	glass	130	~1	ITO target Ar+O_2	$\rho \approx 10^3$ Ωcm $n \approx 10^{21}/cm^3$ $\mu \approx 10$ cm/V.S	18)
		RF-MSP	glass	40	(sputter power) 200W, φ100 target	ITO target Ar 4×10^{-3} Torr	R/□≤10~100 Ω/□	—
$K_3Li_2Nb_5O_{15}$ (KLN)	SC	RF-SP	$K_2BiNb_5O_{15}$ (KBN) sapphire	600~700	0.2	KLN target (K, Li rich)	ferroelectrics $\varepsilon^*=140$, Tc=460°C $n_0=2.277$ (6328 Å)	19)
	a	RF-SP	fuzed quartz	RT	0.38	LNA target Ar/O_2=1	$\varepsilon^* \approx 10^4$ (200~300°C, 1 kHz)	20)

(4)

Materials	Structure *)	Dep. method **)	Substrate	Dep. conditions			Film properties	Ref.
				Sub. temp. (°C)	Dep. rate (µm/hr)	Miscellanea		
$LiNbO_3$ (LN)	SC	RF-SP	c-sapphire	500	0.025	LN target Ar+O_2	ferroelectrics $n_0=2.32$, optical wave guide loss 9 dB/cm (6328 Å)	21)
	SC		c-$LiTaO_3$	850			$n_0=2.288$, optical wave guide loss 11 dB/cm (6328 Å)	22)
$PbTiO_3$	a	DC-MSP	glass	200	0.3	Ti/Pb target Ar/O_2=1	$\varepsilon^* \approx 120$ (RT) Tc≈490°C	23)
	PC	RF-SP	Pt	610	0.34~0.3	PbO/TiO_2 target Ar+O_2	$\varepsilon^* \approx 200$ (RT)	24)
	SC	RF-SP	c-sapphire	620	0.3~0.6	$PbTiO_3$ powder Ar+O_2		—
PZT	PC	EB	fuzed quartz SUS	sputter 350 anneal 700			ferroelectrics $\varepsilon^* \approx 100$ (RT) Ps~4.2 µC/cm^2, Tc~340°C	25)
	PC	RF-SP	fuzed quartz Pt	>500		P2T 52/48 target Ar+O_2	ferroelectrics $\varepsilon^* \approx 751$ (RT) Ps ~21.6 µC/cm^2 Tc ~325°C, $n_0=2.36$	26)

Table 5.2: Deposition methods for compound thin films. (continued)

(5)

Materials	Structure *)	Dep. method **)	Substrate	Dep. conditions			Film properties	Ref.
				Sub. temp. (°C)	Dep. rate (μm/hr)	Miscellanea		
PLT	SC	RF-SP	MgO (100)	600 ~ 700	0.18 ~ 0.48	PLT 18/100 target Ar+O₂	$\epsilon^*\le 700$ $n_0<2.3\sim2.5$ (6328 Å)	27)
PLZT	PC	RF-SP	fuzed quartz Pt	sputter ~ 500 anneal 650 ~ 700	0.2 ~ 0.4	PLZT 7/65/35 target Ar+O₂	$\epsilon^*\approx 1,000\sim1300$ Tc ~170°C	28)
	SC	RF-SP	c-sapphire SrTiO₃ (100)	700	~0.4	PLZT 9/65/35 target Ar+O₂	$n_0<2.49$ (6328 Å)	29)
WO₃	a	VE	glass	100				30)
AlN	PC	RF-SP	glass	200 ~ 300		AlN target Ar	$\rho \sim 2,000$ μΩcm	31)
	PC	DC-MSP	glass	320	1.3	Al target Ar/N₂	c-axis orientation σ=2.9 ~ 5.4°	32)
	SC/PC	RF-MSP	c-sapphire/ glass	50 ~ 500	0.2 ~ 0.8	Al target Ar/N₂	σ=1° (sapphire) σ=3° (glass)	33)
	SC	RF-SP	c, R-sapphire	1200	0.5	Al target NH₃	SAW velocity ~5500 m/s $k^2=0.05\sim0.02\%$	34)

(6)

Materials	Structure *)	Dep. method **)	Substrate	Dep. conditions			Film properties	Ref.
				Sub. temp. (°C)	Dep. rate (μm/hr)	Miscellanea		
	SC	CVD	R-sapphire	1200	3	reaction gas (CH₃)Al+ NH₃+H₂	SAW velocity ~6100 m/s k^2 ~0.8%	35)
CrN	PC	ARE	glass	30 ~ 450		reaction gas NH₃	Ts > 200°C : Cr₂N Ts > 400°C : CrN	36)
BN	PC	ARE	stainless steel glass, Si NaCl	450	9	source H₃BO₃ reaction gas NH₃	Hv ≈ 2128 kgfmm⁻² cubic boron nitride optical gap, 3.64 eV	37)
Si₃N₄		P-CVD	Si	250	3 (200 W)	reaction gas N₂+NH₃+SiH₄	$n_0 \approx 2.0 \sim 2.1$	38)
		RF-MSP	Si	100	1	Si₃N₄ target Ar	$n_0 \approx 2.1$ (6328 Å)	——
TiN	PC	DC-MSP	glass	150	0.6 ~ 1.8	Ti target Ar/N₂=7/3	ρ=250 μΩcm TCR=150 ppm/°C	39)
	PC	RF-MSP	fuzed quartz	500	1.5	TiN target Ar		——

Table 5.2: Deposition methods for compound thin films. (continued)

(7)

Materials	Structure *)	Dep. method **)	Substrate	Sub. temp. (°C)	Dep. rate (μm/hr)	Miscellanea	Film properties	Ref.
							Dep. conditions	
WC	PC	RF-MSD	stainless steel	200 ~ 500	0.36	W target Ar+C_2H_2	Ts ≤ 200°C : WC, W_2C, W_3C mixed, phase Ts ≤ 400 ~ 500°C : WC single phase	40)
				300 ~ 500	0.36 ~ 4.9	W target Ar+C_2H_2	Hv* = 2365 ~ 3200 kgfmm^{-2} friction coefficient ~0.09 * Hv = 3200 kgfmm^{-2} (WC)	
B_4C	a	RF-SP	sapphire	450	~0.5	B_4C target Ar	Hv = 4800 kg/cm^2	41)
CrC	PC	RF-MSP	glass	600	0.2 ~ 0.7	Si target Ar+CH_4	IR absorption, 800 cm^{-1} (Si-C) 2000 cm^{-1} (Si-H)	42)
SiC	PC	RF-SP	glass alumina	550	0.5 ~ 1	SiC target Ar	β-SiC, (220) orientation ρ ≤ 2000 Ωcm B ≤ 2100 K	41) 43)
	SC	IP	Si (111)	1000	0.9 ~ 1.8	reaction gas C_2H_2	β-SiC	44)
	SC	CVD	Si (100)	1330	4 ~ 6	reaction gas H_2+SiH_4 +C_3H_8	β-SiC carbon buffer layer	45)

(8)

Materials	Structure *)	Dep. method **)	Substrate	Sub. temp. (°C)	Dep. rate (μm/hr)	Miscellanea	Film properties	Ref.
							Dep. conditions	
GaAs	SC	MBE	GaAs	600		graphite crucible Ga (1090°C), As (320°C)	n_{300} = 2.0×10^{15}/cm^3 $μ_{300}$ = 7500 cm^2/V.s	46)
	SC	RF-SP	GaAs (100)	500 ~ 625	0.7 ~ 1.2	GaAs target Ar	ρ ≤ 10^5 ~ 10^6 Ωcm $μ_{300}$ ≤ 5000 cm^2/V.s	47)
GaSb	SC	MBE	GaAs				P-type semiconductor n_{300} = 4~6×10^{16}/cm^3 $μ_{300}$ = 670 cm^2/V.s	48)
	a	RF-SP	BaF_2 (111)	400	0.15	GaSb target Ar		49)
InAs	SC	MBE	GaAs	450 ~ 600	0.36 ~ 1		n-type semiconductor n_{300} = 4~6×10^{16}/cm^3 $μ_{300}$ = 1670 cm^2/V.s	48)
$In_{1-x}Ga_xSb$	SC	RF-SP	BaF_2 (111)	400	0.15	InSb, GaSb target Ar	x = 0.36	49)
Nb_3Sn		DC-MSP	sapphire	650 ~ 800	60	Nb_3Sn target Ar	superconducting transition Tc ≤ 18.3 K	50)

Table 5.2: Deposition methods for compound thin films. (continued)

(9)

Materials	Structure *)	Dep. method **)	Substrate	Dep. conditions			Film properties	Ref.
				Sub. temp. (°C)	Dep. rate (μm/hr)	Miscellanea		
Al$_5$ Nb$_3$Ge	PC	SP	sapphire	700	0.1	sputter gas pressure 40 Pa	Tc = 21.6 K ΔTc = 3.4	51)
	PC	CVD	Nb tape	900	216	NbCl$_5$ GeCl$_4$ H$_2$	Tc = 20.1	52)
B1 NbN	PC	SP	sapphire	<350 anneal 600			Tc = 16 K H$_{c2}$ = 28 T (13 K) J$_C$ = 8×10^5 A/cm^2 (0 T, 4.2 K)	53)
	PC	CVD	carbon fiber	1400 ∿ 1600	7 ∿ 11	NbCl$_5$ N$_2$ H$_2$	Tc = 16.4 K H$_{c2}$ = 11 T J$_C$ = 10^6 A/cm^2 (0 T, 4.2 K)	54)
	PC	P-CVD	carbon fiber	1100	3.6 ∿ 7	NbCl$_5$ CH$_4$ NH$_3$	Tc = 15 ∿ 17 K H$_{c2}$ = 21 T (4.2 K) J$_C$ = 10^6 A/cm^2 (0 T, 4.2 K)	55)
PbMo$_6$S$_8$	PC	SP		anneal 750 ∿ 1100		MoS$_2$ Mo PbS	Tc = 14 K J$_C$ = 8×10^4 A/cm^2 (0 T, 4.2 K)	56)
	PC	SP	sapphire	anneal 1000		Mo sputter Pb/MoS$_2$ 1000°C	Tc = 14 K J$_C$ = 10^6 A/cm^2 (0 T, 4.2 K)	57)

(10)

Materials	Structure *)	Dep. method **)	Substrate	Dep. conditions			Film properties	Ref.
				Sub. temp. (°C)	Dep. rate (μm/hr)	Miscellanea		
BaPb$_{1-x}$Bi$_x$O$_3$	PC	SP	glass	up to 404		anneal 550°C 12 hr in Air, O$_2$	Tc = 9 K	58)
	SC	MSP	(100) SrTiO$_3$	700		Ar/O$_2$=1	Tc = 10.5 K ΔTc ≪ 1 K x = 0.3 M$_H$ = 55 cm^2/V.s x = 0.2 = 0.35	59)
LSC K$_2$NiF$_4$	PC	SP	sapphire MgO	660 820	0.8	Ar Ar/O$_2$	Tc ≤ 40 K (001) orientation T$_{R=0}$ ≅ 24 K crystalization, Ts > 450°C	60)
	SC (001)	M-SP	(100) SrTiO$_2$	800		n=6.28×10^{21} cm^{-3} μ$_H$=1.3 cm^2/V.s ρ∿25μΩcm (Tc)	Tc = 14 K (as sp.) 22 K (anneal) post anneal 800°C 8 hr	61)
	PC	SP	YSZ	400 ∿ 450	0.125	target co-sputter LSC> LBC LSC > LCaC	Tc = 45 K T$_{R=0}$ ≅ 17 K Jc = 150 A/cm^2 (0 T, 4.2 K) post anneal in O$_2$ 760 ∿ 850°C, 15 hr	62)

Table 5.2: Deposition methods for compound thin films. (continued)

(11)

Materials	Structure *)	Dep. method **)	Substrate	Dep. conditions			Film properties	Ref.
				Sub. temp. (°C)	Dep. rate (μm/hr)	Miscellanea		
LSC K_2NiF_4	SC (001)	MSP	(100) $SrTiO_3$	600	0.6	Ar	$T_c \approx 34$ K $\Delta T \sim 3$ K $T_{R=0} = 25$ K $J_c \sim 30$ A/cm² (0 T, 4.2 K) anneal 900°C, 72 hr	63)
	PC	VE (EB)	sapphire			La_2O_3, Sr, Ca O_2 RVD (1×10^{-4} Torr) anneal	$T_c > 30$ K $T_{R=0} = 15$ K energy gap 20 ~ 30 mV	64)
	PC	VE (EB)	MgO sapphire (100) $SrTiO_3$	450		anneal 900 ~ 950°C in O_2	$T_c \approx 97$ K $T_{R=0} \approx 87$ K $J_c = 7\times10^5$ A/cm² (0 T, 77 K) $= 5\times10^7$ A/cm² (0 T, 4.2 K)	65)
	PC	EB	(100) $SrTiO_3$	200 ~ 870°C	1 ~ 4.3 0.5 ~ 1 μm	O_2 RVD anneal 650 ~ 850°C	10% ~ 90% R-transition 88 K $\rho \sim 0.5$ mΩcm (Tc) a-axis orientation (a >> c) $J_c{/\!/} = 2\times10^6$ A/cm² (0 T, 4.2 K) $= 9\times10^4$ A/cm² (0 T, 78 K)	66)

(12)

Materials	Structure *)	Dep. method **)	Substrate	Dep. conditions			Film properties	Ref.
				Sub. temp. (°C)	Dep. rate (μm/hr)	Miscellanea		
YBC $YBa_2Cu_3O_{7-\delta}$	SC	MSP	(110) $SrTiO_3$	700		sputter Ar/O=1 8×10^{-2} Torr anneal in O_2 920°C, 2 hr	$T_{R=0} \approx 84$ K, $\Delta T = 6$ K $J_c{/\!/} = 1.8\times10^6$ A/cm² (0 T, 77 K) $J_c\perp = 3.2\times10^4$ A/cm² (0 T, 4.2 K) $\rho{/\!/} = 0.5$, $\rho\perp = 16$ mΩcm (Tc)	67)
	PC	MSP	R-sapphire	200	0.9	sputter, Ar anneal	$T_{R=0} \approx 70$ K, $T_{c0} = 94$ K	68)
	SC	MSP	(100) MgO	650	thickness 1 μm	in situ dep.	$T_{R=0} \approx 86$ K, $\Delta T = 3$ K $\rho = 0.5$ mΩcm (Tc) Er-Ba-Cu-O	69)
	PC	VE (KrF excimer laser)	(100) $SrTiO_3$ R-sapphire	450	1 Å/pulse KrF 1 J/pulse	anneal in O_2 900°C	$T_c \approx 95$ K $T_{R=0} \approx 85$ K ($SrTiO_3$) $\Delta T = 2$ K $T_{R=0} \approx 75$ K (sapphire) $\Delta T = 12$ K	70)
	PC	EB	sapphire YSZ	RT	1 ~ 1.8	Y, Ba, Cu multi-layer anneal in O_2 800 ~ 850°C	$T_c \approx 94$ K $T_{R=0} \approx 72$ K (YSZ) $T_{R=0} \approx 40$ K (sapphire)	71)

Table 5.2: Deposition methods for compound thin films. (continued)

(13)

Materials	Structure *)	Dep. method **)	Substrate	Sub. temp. (°C)	Dep. rate (μm/hr)	Miscellanea	Film properties	Ref.
	PC	MSP	(100) MgO (100) SrTiO₃	700	0.48	anneal in O₂ 800 ~ 900°C	Tc ≈ 115 K, $T_{R=0}$ = 104 K, ρ = 50 μΩcm (Tc), Jc > 20×14⁴ A/cm² (0 T, 77 K)	72)
	PC	MSP co-sputter	(100) MgO	~RT	thickness: 0.4 μm	target: Bi, SrCa, Cu anneal in O₂ 865°C	C-axis orientation $T_{R=0}$ ~ 80 K, $B_{c2}^{\perp}$ ~ 8.5 T/K, $B_{c2}^{\parallel}$ ~ 0.56 T/K	73)
BSCC Bi-Sr-Ca-Cu-O	PC	EB	(100) MgO	~RT	0.2 μm	source: Bi, SrF₂, CaF₂, Cu anneal in O₂ 860 ~ 890°C	c-axis orientation c = 30.605 A, Tc ~ 110 K, $T_{R=0}$ ~ 60 K	74)
	PC	EB	(110) (100) SrTiO₃	~RT	thickness: 0.3 ~ 0.5 μm	source: Bi, SrF₂, CaF₂, Cu anneal in wet O₂ 850°C, 5 min	c-axis orientation, epitaxial $T_{R=0}$ ~ 80 K, Jc = 1×10⁶ A/cm² (0 T, 4.2 K), ρ = 150 μΩcm (Tc0)	75)
	PC	LA (ArF eximer laser)	(100) MgO	RT	thickness: 0.1 ~ 2 μm	target BSCCO eximer laser, 10 J/cm². anneal in air, 890°C 1 min.	c-axis orientation Tc ~ 80 K (c = 30.8 Å) Tc ~ 120 K (c = 36.8 Å)	76)

(14)

Materials	Structure *)	Dep. method **)	Substrate	Sub. temp. (°C)	Dep. rate (μm/hr)	Miscellanea	Film properties	Ref.
	PC	IBS	(100) MgO ZrO₂	~RT	thickness: 0.3 ~ 1 μm	anneal in O₂ 825 - 880°C	Bi₂.₀₀Sr₂.₀₉Ca₀.₇₅⁻Cu₂.₄₄Oₓ, Tc ~ 85 K, $T_{R=0}$ ~ 75 K Bi₄.₀₀Sr₃.₁₂Ca₃.₀₃⁻Cu₄.₅₆Oₓ, Tc ~ 110 K, $T_{R=0}$ ~ 63-67 K	77)
BSCC Bi-Sr-Ca-Cu-O	PC	MSP layer-by-layer dep. multi-target	(100) MgO	650°C	~ 100 Å/hr thickness: 200 - 400 Å	target: Bi, SrCu, CaCu anneal in O₂ 855°C, 5 hr.	c-axis orientation Bi₂Sr₂Ca₁Cu₂Oₓ Tc0 ~ 80 K, Bi₂Sr₂Ca₂Cu₃Oₓ Tc0 ~ 110 K, Bi₂Sr₂Ca₃Cu₄Oₓ Tc0 ~ 90 K	78)
	PC	CVD	(100) MgO	910°C	8 thickness, 8 μm	source: bismuth alkoxide, β-diketonate chelates of Sr, Ca, Cu.	c-axis orientation, $T_{R=0}$ ~ 78 K	79)
TBCC Tl-Ba-Ca-Cu-O system	PC	EB	YSZ	<50°C	thickness: 0.7 μm	source: Tl, Ba, Ca, Cu. dep. in O₂. anneal in air 850°C, 5 min.	Tc ~ 110 K (Meissner) $T_{R=0}$ ~ 97 K, Jc ~ 7.1×10⁵ A/cm² (0 T, 76 K) Tl₂Ba₂Ca₁Cu₂Oₓ	80)

Table 5.2: Deposition methods for compound thin films. (continued)

(15)

Materials	Structure *)	Dep. method **)	Substrate	Dep. conditions			Film properties	Ref.
				Sub. temp. (°C)	Dep. rate (μm/hr)	Miscellanea		
TBCC Tl-Ba-Ca-Cu-O system	PC	MSP	YSZ	∿RT	thickness 0.4 μm	target: Tl, BaCa, Cu. anneal in O_2 850°C, 4 min.	c-axis orientation Tc ∿ 110 K $T_{R=0}$ ∿ 96 K Jc ∿ 2×10^4 A/cm^2 (0 T, 4.2 K)	81)
	PC	SP	(100) MgO SrTiO₃ YSZ	∿RT	0.48 thickness 1 ∿ 4 μm	anneal in O_2/Tl, 880 - 900°C in sealed quartz	c-axis orientation. $Tl_2Ca_1Ba_2Cu_2O_x$ $T_{R=0}$ ∿ 100 K $Tl_2Ca_2Ba_2Cu_3O_x$ $T_{R=0}$ ∿ 110 K $Tl_2Ca_2Ba_3Cu_4O_x$ $T_{R=0}$ ∿ 120 K Jc ∿ 1.5×10^5 A/cm^2 (77 K)	82)
	PC	SP	(100) MgO SrTiO₃	RT	0.2 thickness 0.2∿1.0 μm	anneal in O_2 or air 800 - 880°C	c-axis orientation. Rocking curve width 0.32°. $Tl_2Ba_2Ca_1Cu_2O_8$ T_{c0} ∿ 102 K Jc ∿ 10^4 A/cm^2 (90 K). $Tl_2Ba_2Ca_2Cu_3O_{10}$ T_{c0} ∿ 116 K Jc ∿ 10^5 A/cm^2 (100 K)	83)
	PC	MSP	(100) MgO	200°C	0.6	anneal in O_2/Tl 850 ∿ 900°C	c-axis orientation. $Tl_2Ba_2Ca_1Cu_2O_x$ $T_{R=0}$ ∿ 102 K Jc ∿ 1.2×10^5 A/cm^2 (77 K) $Tl_2Ba_2Ca_2Cu_3O_x$ $T_{R=0}$ ∿ 117 K $Tl_1Ba_2Ca_3Cu_4O_x$ $T_{R=0}$ ∿ 113 K	84) I 86)

Table 5.2: Deposition methods for compound thin films. (continued)

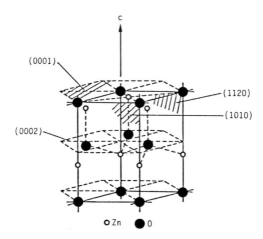

Figure 5.1: Crystal structure of ZnO single crystal.

Crystal system	6 mm (wurtzite)
Space group	$P6_3$ mc
Lattice constant	a = 3.24265 Å, c = 5.1948 Å
Sublimation point	1975 ± 25 °C
Hardness	4 Mohs
Dielectric constants	ε_{11}^S = 8.55, ε_{33}^S = 10.20 × 10^{-11} F/m
Density	5.665×10^3 kg/m^3
Thermal expansion coefficient	α_{11} = 4.0, α_{33} = 2.1 (×10^{-6} /°C)
Optical transparency	0.4 - 2.5 μm
Refractive index	n_o = 1.9985, n_e = 2.0147 (λ = 6328 Å)
Electro-optic constant	r_{33} = 2.6, r_{13} = 1.4 (×10^{-12}m/V, λ = 6328 Å)

Table 5.3: Physical properties of ZnO single crystal.

Owing to these excellent piezoelectric properties, thin films of ZnO are used for making ultrasonic transducers in high frequency regions. In the past 15 years, many workers have investigated fabrication processes for ZnO, including sputter deposition (90), chemical vapor deposition (91), and ion plating (92). Among these processes sputter deposition is the most popular process for ZnO thin film deposition.

5.1.1.1 Deposition of ZnO

Polycrystalline thin films: Polycrystalline ZnO thin films with a c-axis orientation are one of the most popular piezoelectric thin films. A typical sputtering system for the deposition of the c-axis oriented ZnO films is shown in Fig. 5.2. The basic system consists of a planar diode sputtering device. Zn metal or ZnO ceramic is used as the cathode (target). Sputtering is done with a mixed gas of Ar and O_2. A dc high voltage is supplied for the Zn metal or ZnO ceramic target. The electrical conductivity of the ZnO ceramic target should be higher than 10^3 mho in order to keep the dc glow discharge between the electrodes. When rf high voltage is used for sputtering, i.e. rf-diode sputtering, a high resistance ZnO target can be used.

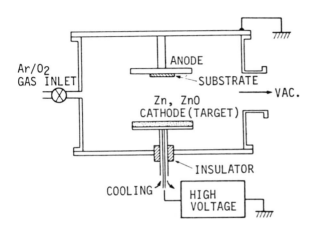

Figure 5.2: Typical sputtering systems for the deposition of ZnO thin films.

The ZnO ceramic target for rf-sputtering is prepared as follows; first, ZnO powder (purity > 99.3%)is sintered in air at 800 to 850°C for one hour. The sintered powder is then pressed at about 100 kg/cm² into the form of the target (typically a disk) and is finally sintered at 930°C for two hours. The resultant ZnO ceramic is not completely sintered until the ceramic is mechanically shaped into the final form of the target. The high conductivity ZnO ceramic for dc sputtering is made by sintering of the ZnO at a higher temperature of around 1300°C.

Polycrystalline ZnO films with a c-axis orientation are commonly deposited on a glass substrate by dc or rf-sputtering when the temperature of the substrate is kept at 100 to 200°C. Figure 5.3 shows their reflection electron diffraction patterns. A typical cross section observed by SEM is shown in Fig. 5.4. The c-axis oriented films consist of a so-called fiber structure and the c-axis is oriented normally to the substrate plane.

In general, the c-axis orientation is frequently observed in these deposited films. This is reasonably well understood since the c-plane of the ZnO crystallites corresponds to the densest packed plane, and the growth mechanism of the sputtered ZnO thin films will be governed by Bravais' empirical law for crystal growth.

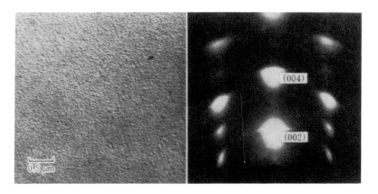

Figure 5.3: Surface structure and RED pattern of ZnO film of c-axis orientation about 2 μm thick prepared on glass substrate.

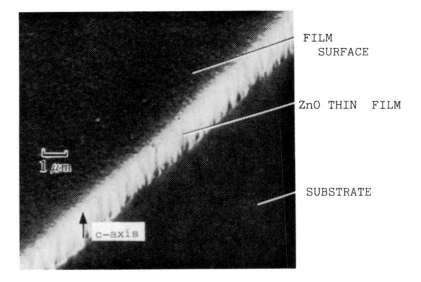

Figure 5.4: SEM image showing cleaved section of ZnO film of c-axis orientation about 2 μm thick prepared on glass substrate.

The reflection electron diffraction patterns of ZnO films of different film thickness on the glass substrate are shown in Fig. 5.5 (93). The (002) orientation is clearly observed at film thickness of more than 500Å. The angular spread of (002) arcs decreases with film thickness. The typical half-angular spread was within $7.5 - 9°$ for the films of $0.5 - 10\mu m$ thick. The mean inclination of the c-axis from the substrate normal was within $3°$. The lattice constant c_0 of the film is 5.23 to 5.24Å. The c_0 is slightly longer than the bulk single-crystal value ($c_0 = 5.2066$Å, $a_0 = 3.2497$Å)(94).

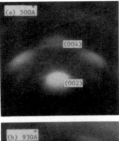

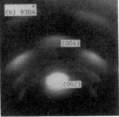

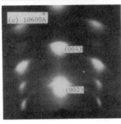

Figure 5.5: Typical electron diffraction patterns of sputtered ZnO thin films on glass substrates for various film thickness: (a) 500Å, (b) 930Å, (c) 10,600Å.

Sputtering parameters: The detailed studies of ZnO thin film growth suggest that the structure of sputtered ZnO films depends on various sputtering parameters, including deposition rate, substrate temperature, gas pressure and composition, residual gas, and target composition. In some cases the growth of ZnO films does not obey Bravais' empirical law and do not show a c-axis orientation.

Deposition rate/substrate temperature: The deposition rate and the substrate temperature will drastically influence the crystal structure of deposited films. Figure 5.6 shows the optimum condition for deposition of c-axis oriented films by various sputtering systems (94). It shows that the optimum condition of the substrate temperature is 100 to 200°C for a deposition rate below 1μm/hr with the rf-diode sputtering system. In the magnetron sputtering system, the optimum condition shifts to a higher substrate temperature with a higher deposition rate; typically the substrate temperature is 300 to 400°C; the deposition rate, 1 to 5μm/hr (95).

A high deposition rate with a low substrate temperature and/or a low deposition rate with a high substrate temperature frequently causes mixed orientation of the c-axis and the a-axis in sputtered films. A typical RED pattern of the ZnO films with the mixed orientation is shown in Fig. 5.7.

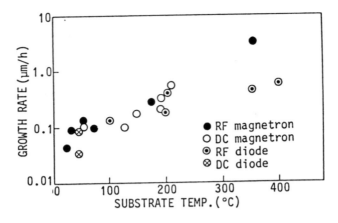

Figure 5.6: Optimum sputtering conditions for the deposition of c-axis oriented film on a glass substrate for various sputtering systems.

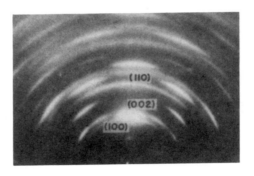

Figure 5.7: Electron diffraction pattern of ZnO thin film showing mixed orientation.

  Sputtering gas composition/gas pressure: Westwood reported that an optimum partial oxygen pressure existed in the mixed gas of (Ar + O$_2$) for making a c-axis oriented ZnO films (96). The optimum content of O$_2$ was reported to be 67% at a total sputtering gas pressure of 3.5×10^{-2} Torr. The pressure of the optimum oxygen pressure suggests that favorable oxidization at the surface of the cathode target and/or deposited films is necessary for obtaining the c-axis oriented ZnO films. When the sputtering gas pressure is reduced below 1×10^{-3} Torr, a different feature is observed (97). The c-axis orientation is suppressed.

  Table 5.4 shows the crystallographic orientation for different gas pressures. Two types of orientation are observed. One is the c-axis orientation (normal orientation, c$_\perp$). The other is the c-axis parallel to the substrate (parallel orientation, c$_\parallel$) in this case either <110> or <100> axis is normal to the film surface. Figure 5.8 gives a composite plot of the crystallographic orientation, substrate temperature, and deposition rate for the ZnO sputtered films at low gas pressure (1×10^{-3} Torr), showing that the parallel orientation is predominant. In contrast the normal orientation is predominant at high gas pressure (greater than 3×10^{-2} Torr).

Sample No.	Total sputtering pressure* (10^{-3} torr)	Substrate temperature (°C)	Condensation rate (µm/h)	Thickness (µm)	Crystal- lographic orientation
1-0	1	40	0.03	0.1	C⊥
1-1	1	40	0.1	0.25	C⊥
1-2	1	40	0.12	0.36	C⊥, C∥
1-3	1	150	0.1	0.3	C∥
1-4	1	150	0.7	0.3	C∥
1-5	1	200	0.7	0.3	C∥
1-6	1	270	0.07	0.2	C∥
1-7	1	270	0.6	0.3	C∥
1-8	5	200	0.2	0.1	C∥
1-9	30	100	0.15	0.15	C⊥
1-10	30	200	0.15	0.15	C⊥
1-11	60	200	0.2	0.2	C⊥
1-12	100	200	0.15	0.15	C⊥

* Al + O_2, 50% O_2

Table 5.4: Crystallographic orientation of ZnO thin films for different sputtering gas pressure.

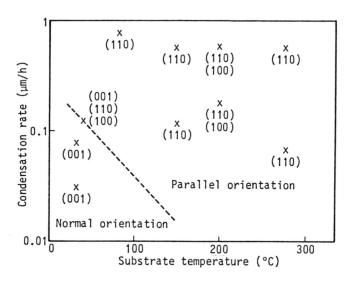

Figure 5.8: Variation of crystallographic orientation for ZnO thin films with deposition rate and substrate temperature sputtered in low gas pressure (1×10^{-3} Torr).

The crystalline properties of the resultant films are influenced by the degree of oxidation of both surface of target and film during deposition. The degree of oxidation at the film surface increases with increasing substrate temperature. Under low oxygen gas partial pressure a high substrate temperature is necessary for obtaining a favorable degree of oxidation. Figure 5.9 shows the optimum sputtering conditions for deposition of c-axis oriented films at various gas pressures and substrate temperatures. It stands to reason that lowering the gas pressure shifts the optimum substrate temperature to higher values.

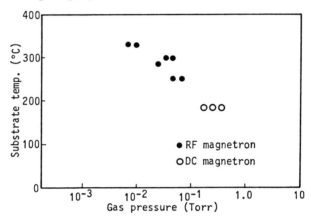

Figure 5.9: Optimum sputtering conditions for the deposition of c-axis oriented ZnO films for various sputtering gas pressure.

Target composition/impurity: A sintered ZnO target is more suitable for making c-axis oriented films than a Zn metal target. Foster pointed out that the presence of an organic vapor, such as methane, in the sputtering atmosphere reduced the growth of c-axis and induced the a-axis orientation (98).

It is also interesting to note that the addition of foreign atoms, such as aluminum and copper, during film growth changes the crystallographic orientation of the sputtered ZnO films. Table 5.5 indicates the typical change of orientation with aluminum and copper. The proportions of aluminum and copper were determined by chemical analyses. Typical photographs (electron micrographs) and reflection electron diffraction patterns are shown in Fig. 5.10. The results are summarized as follows: the addition of aluminum enhances the growth of parallel orientation (c-axis parallel to the film surface, $c_{||}$), and the increase of aluminum results in an increase of parallel orientation intensity. For a favorable amount of aluminum the normal orientation (c-axis normal to the film surface, $c_{\perp}$) disappears and the parallel orientation remains.

Copper has the opposite effect on crystallographic orientation. It enhances the growth of normal orientation.

Electrode configuration/substrate position: The electrode configuration of the sputtering system also affects the crystallographic orientation. Two types of rf-sputtering systems are shown in Fig. 5.11. One is a conventional planar electrode system, Fig. 5.11 (a), the other is a hemispherical electrode system, Fig. 5.11 (b). The glass substrates are placed on a substrate holder behind the anode. The thin films of ZnO with a c-axis orientation are deposited under the conditions shown in Fig. 5.6. However, the electrode configuration affects the degree of the crystalline orientation.

Sample No.	Total sputtering pressure* (10^{-3} torr)	Contents of foreign metals (at. %)	Substrate temperature (°C)	Condensation rate (µm/h)	Thickness (µm)	Crystallographic orientation
2-0	1	–	40	0.1	0.3	c⊥
2-1		6 (Al)				c∥
2-2		–				c∥
2-3	1	0.1 (Al)	200	1.2	0.6	c∥
2-4		1.3 (Al)				c∥
2-5		6 (Al)				c∥
2-6		–				c⊥
2-7	30	0.04 (Al)	200	0.075	0.15	c⊥ + c∥
2-8		5 (Al)				c∥
2-9		–				c∥
2-10	1	0.013 (Cu)	200	0.9	0.45	c∥
2-11		0.5 (Cu)				c⊥ + c∥

* Al + O_2, 50% O_2

Table 5.5: Crystallographic orientation of ZnO thin films with admixed foreign atoms.

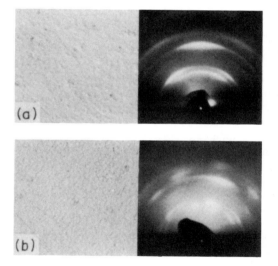

Figure 5.10: Electron micrographs of ZnO thin films of 0.3µm thick with admixed foreign atoms: (a) with Al of 6 atm.%, (b) with Cu of 0.5 atm.%.

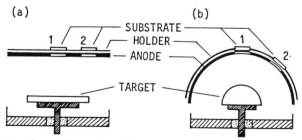

Figure 5.11: Electrode configuration of two sputtering experiments: (a) conventional planar system and (b) hemispherical system.

Figure 5.12 show typical x-ray diffraction patterns from the ZnO thin films on glass substrates prepared by the two sputtering systems. It shows that ZnO thin films prepared by the planar system at substrate position "1" exhibits very poor c-axis orientation.

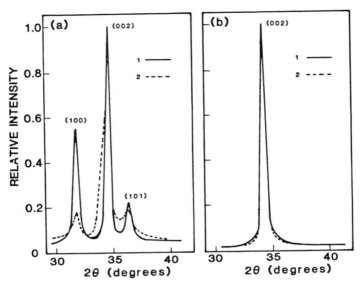

Figure 5.12: X-ray diffraction patterns from ZnO thin films on glass substrates prepared by (a) planar system and (b) by hemispherical system.

The degree of c-axis orientation of the film prepared at position "2", increases somewhat but (100) and (101) peaks still remain in the diffraction pattern. Such variation of the c-axis orientation with the substrate position is also observed in dc-sputtered ZnO thin films (99). In addition it has been found that the degree of the c-axis orientation is strongly affected by a slight change of sputtering conditions. On the other hand, the ZnO thin films prepared by the hemispherical system normally exhibit excellent c-axis orientation regardless of the substrate position as shown in Fig. 5.11 (b). From an x-ray rocking-curve analysis, the standard deviation of c-axis orientation is found to be less than 3° for ZnO thin films prepared by the hemispherical system. In the hemispherical system, the distribution of incident angles of the sputtered particles at the substrate surface is considered to be much narrower than in the planar system. This probably causes the formation of a beam-like flow of the sputtered particles, including Zn atoms, onto the substrate resulting in the growth of highly oriented ZnO thin films.

Since the hemispherical system shows uniformity in film thickness and crystal orientation, the system is useful for the production of ZnO thin films. Figure 5.13 shows the construction of an rf-sputtering system designed for production. In this system, diameters of the ZnO target and substrate holder (anode) are 70 mm and 220 mm, respectively. Sixty substrate wafers of 25 mm square can be loaded on the substrate holder. The holder is rotated for further improvement in uniformity of film thickness. The sputtering currents are controlled so as to flow uniformly at the whole target surface, then the thickness variation in a single wafer $(h_o - h_r)/h_o$ is simply given by the relationship;

$$(h_o - h_r)/h_o \approx (3/2)\,(r/L)^2, \qquad (5.1)$$

where h_o is the thickness at a center of the wafer, h_r is the thickness at distance r apart from the center of the wafer, and L is the space between the wafer and the target center. Taking, L = 11 mm, r = 12.5 mm, the maximum thickness variation in the wafer is ± 1 %. The thickness variation between different substrate wafers is less than 1 % which is governed by the geometrical accuracy of the electrode system. The thickness of the ZnO thin films is monitored by a laser interference device during deposition. A photograph of the sputtering chamber with the ZnO target in shown in Fig. 5.14.

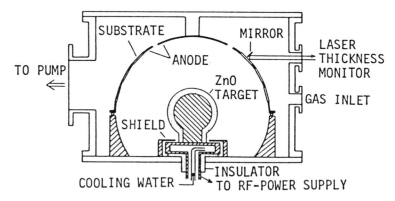

Figure 5.13: Schematic diagram of the hemispherical sputtering system for ZnO thin film deposition.

Figure 5.14: Photograph of the sputtering chamber of the hemispherical system with ZnO target.

Ion plating, cluster ion beam deposition, and chemical vapor deposition are also used for the deposition of ZnO thin films. The difference of the growth mechanism of ZnO thin films in these deposition processes is not well understood yet. However, it is very interesting that the ion plating process shows nearly the same optimum condition as the sputtering process in growth rate and substrate temperature indicated in Fig. 5.9 (92). This suggests that the growth mechanism of the ZnO thin films in the ion plating process is resembles that of the sputtering process. Recent experiments suggest that the ECR plasma CVD process can also be applied for deposition of c-axis oriented ZnO thin films (100).

Single crystal films: Two types of piezoelectric ZnO single crystal films are epitaxially grown on sapphire single crystal substrates by rf-sputtering (101). Epitaxial relations between these ZnO thin films and the sapphire (Al_2O_3) substrates are determined as follows:

$$(0001)ZnO \parallel (0001)Al_2O_3 , [11\bar{2}0]ZnO \parallel [10\bar{1}0]Al_2O_3,$$

$$(5.2)$$

$$(11\bar{2}0)ZnO \parallel (01\bar{1}2)Al_2O_3 , [0001]ZnO \parallel [0\bar{1}11]Al_2O_3.$$

A ZnO target is used for sputter deposition of the epitaxial films as well as deposition of the polycrystalline films. The key sputtering parameters for making epitaxial films are the deposition rate and the substrate temperature. Figure 5.15 shows typical variations of their crystal properties with the sputtering conditions. Epitaxial single crystal films are grown at a substrate temperature of 400 to 600°C. Typical electron micrographs with electron diffraction patterns for these ZnO films are shown in Fig. 5.16. The epitaxial films are very smooth and no texture is observed. The lattice constant c_0 of the film on (0001) sapphire is 5.210Å and a_0 of the film on (0112) sapphire is 3.264Å. The c_0 of the film is almost equal to the standard (+0.065%), but a_0 is greater than the standard (+0.44%). This phenomenon can be explained by stress resulting from the difference of thermal expansion characteristics of film and substrates. The thermal expansion coefficients across the c-axis for ZnO and sapphire show almost the same values: 5.5ppm/°C for ZnO and 5.42 ppm/°C for sapphire. Therefore little stress is induced through cooling down after the deposition for the (0001)/ZnO/(0001)Al_2O_3 structure, where the c-axis of both ZnO and sapphire are perpendicular to the surface. On the other hand, the thermal expansion coefficient along the c-axis for ZnO is considerably smaller than that of the sapphire; ZnO shows 3.8ppm/°C and sapphire shows 6.58ppm/°C. In the (11$\bar{2}$0)ZnO/(01$\bar{1}$2)Al$_3$O$_3$ structure, the c-axis of ZnO is parallel to the surface, and that of sapphire is nearly parallel to the surface. This suggests that the compressive stress parallel to the surface is likely to be present in the film after deposition.

The electrical resistivity of these epitaxial ZnO thin films is as low as 10^2 to 10^3Ωcm. Strong piezoelectric properties with high electrical resistivity are observed when the sputter deposition is conducted from Li-doped ZnO targets. The Li-doped ZnO targets are prepared by the addition of Li_2CO_3 before sintering of ZnO to a level of 0.5 to 2 mol %. As-sputtered ZnO thin films doped with Li show the resistivity of around 10^3 to 10^4 Ωcm. The resistivity increases by two orders after post-annealing in air at

600°C for 30 min. Figure 5.17 shows the variation of the resistivity with concentration of Li_2O_3 in the target.

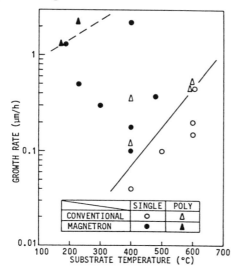

Figure 5.15: Crystalline structures of ZnO films sputtered at various conditions.

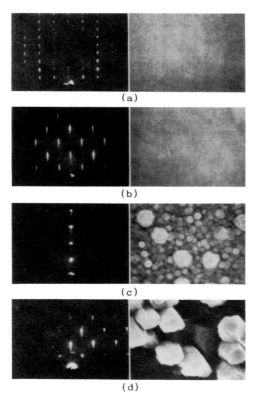

Figure 5.16 Electron micrographs and electron diffraction patterns of ZnO films on sapphire sputtered at a substrate temperature of 600°C with various deposition rates. The orientation of the sapphire substrates and deposition rates are (a) (0001) and 0.2μm/h, (b) (01$\bar{1}$2) and 0.2 μm/h, (c) (0001) and 0.4μm/h, and (d) (01$\bar{1}$2) and 0.44μm/h.

1 μm

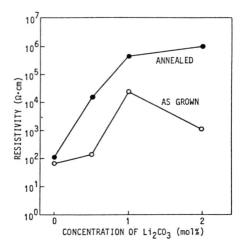

Figure 5.17: Variation of the resistivity of the sputtered ZnO thin films with concentration of Li_2CO_3 in the sputtering target.

The surface flatness of these epitaxial films strongly depends on the growth temperature. Lowering the growth temperature results in a smooth surface. Detailed studies on the epitaxial growth of ZnO single crystal films by sputtering suggests that rf-magnetron sputtering allow a decrease in the growth temperature and improve the surface smoothness of the epitaxial ZnO thin films. Typical sputtering conditions for ZnO single crystal films are listed in Table 5.6. The growth temperature of ZnO single crystal films in magnetron sputtering is as low as $200°C(102)$. The most popular technique for preparation of single crystal ZnO thin films is sputtering, although chemical vapor deposition is also used (103). In chemical vapor deposition, oxidation of Zn vapor or decomposition of zinc oxides or zinc halides are used for the reaction. The epitaxial temperature is around 650 to 850°C. The temperature is lowered in plasma enhanced chemical vapor deposition. The resistivity of the epitaxial films is then as low as $1\Omega cm$. The crystal properties and their surface smoothness are improved by the deposition of a thin ZnO sputtered layer on the substrate as buffer layer (104).

	RF-diode system	RF-magnetron system
Target dimension	30 mmϕ	85 mmϕ
Target-substrate spacing	25 mm	50 mm
Sputtering gas	Ar + O_2 (1:1)	Ar + O_2 (1:1)
Gas pressure	6.7 Pa	0.67 - 1.3 Pa
Rf power	1 - 10 W	25 - 200 W
Substrate temperature	400 - 600 °C	180 - 480 °C
Growth rate	0.04 - 0.5 μm/h	0.1 - 2.2 μm/h

Table 5.6: Typical sputtering conditions for the preparation of ZnO single crystal films on sapphire.

5.1.1.2 Electrical Properties and Applications:

Longitudinal/shear mode couplings: The piezoelectric properties of ZnO thin films are evaluated by measuring the frequency dependence of electrical admittance in a ZnO thin film transducer. The construction of the ZnO thin film transducer and typical admittance characteristics are shown in Fig. 5.18. The transducer is composed of the c-axis oriented ZnO film of about $5\mu m$ thickness sputtered onto the end of a fused-quartz rod (15 mm long and 5 mm in diameter) which is precoated with a thin Cr/Al base electrode. A counter electrode of gold thin film (1 mm in diameter, $0.2\mu m$ thick) is deposited in vacuum onto the ZnO thin film.

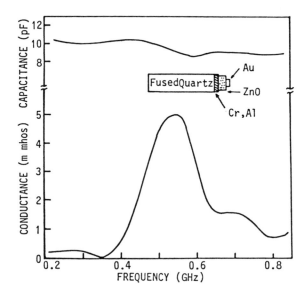

Figure 5.18:   Frequency dependence   of   the admittance   characteristics of sputtered ZnO thin film transducer; ZnO film thickness, about $5\mu m$.

The phase velocity of bulk acoustic wave of longitudinal-mode is calculated from the relationship $v_p = 2f_r d$ where f_r is a frequency at the resonance and d is the ZnO film thickness. The longitudinal-mode electromechanical coupling factor k_t is calculated from the admittance characteristics using the equation.

$$k_t \simeq \pi G_A X_C Z_M / 4 Z_T, \tag{5.3}$$

where G_A is the conductance above the background at antiresonance, and X_C is the transducer capacitive reactance also at the antiresonant frequency. The acoustic impedance of the fused quartz propagation medium and ZnO transducer material are, respectively $Z_M = 1.58 \times 10^7 kg/m^2$ sec and $Z_T = 3.64 \times 10^7 kg/m^2$ sec The admittance is measured by using a network analyzer. The results show that the coupling factors k_t for ZnO thin films with c-axis orientation is $k_t = 0.23$ to 0.24. These coupling factors are 85 to 88% of the bulk single crystal value (105).

The shear mode coupling of the ZnO thin films is evaluated by measuring the frequency dependence of electrical admittance in a ZnO thin film transducer which is composed of a ZnO thin film with parallel orientation.

The ZnO thin films with parallel orientation are prepared by the addition of aluminum during sputtering (106). Figure 5.19 shows a typical frequency response of this kind of transducer measured at unmatched and untuned conditions. Two peaks (A) and (B) are observed. Peak (A) is a resonance point which corresponds to the excitation of shear mode elastic waves. Peak (B) may be the third higher overtone of peak (A).

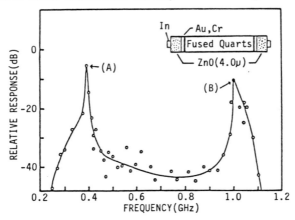

Figure 5.19: Frequency response of two port ZnO thin film acoustic element with parallel orientation.

Surface acoustic wave (SAW) properties: In the past fifteen years many workers have investigated the surface acoustic wave (SAW) properties of ZnO thin films since they are promising materials for making SAW devices for consumer electronics, communication systems, data processing systems, and acoustooptic devices (107). Thin film SAW devices are essentially composed of a layered structure; substrates overcoated by thin piezoelectric ZnO films. Their SAW properties which include phase velocity, electromechanical coupling, and propagation loss, are governed by the thickness of ZnO films, the wavelength of SAW, and the materials constants of ZnO and the substrates.

As shown in Fig. 5.20 there are four types of electrode configurations for the excitation of SAW in a layered structure in which c-axis oriented ZnO films are deposited on the glass substrate, and epitaxial single crystal ZnO thin films are grown on the single crystal substrate.

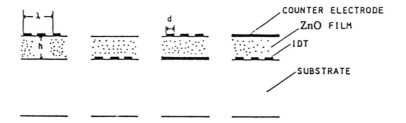

Figure 5.20: Four different configurations of interdigital transducers for the excitation of SAW.

Figure 5.21 shows the calculated values of the phase velocity and electromechanical coupling k^2 for a layered structure. The effective coupling k^2 varies with the ZnO film thickness to wavelength ratio (h/λ). The variations show a double-peaked character for the ZnO/glass structure where the first peak is at $h/\lambda \approx 0.02 - 0.03$ and the second peak at $h/\lambda \approx 0.5(108)$.

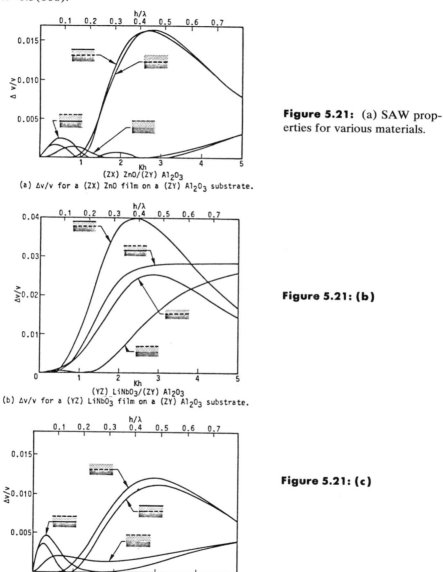

Figure 5.21: (a) SAW properties for various materials.

(a) $\Delta v/v$ for a (ZX) ZnO film on a (ZY) Al_2O_3 substrate.

Figure 5.21: (b)

(b) $\Delta v/v$ for a (YZ) LiNbO$_3$ film on a (ZY) Al_2O_3 substrate.

Figure 5.21: (c)

(c) $\Delta v/v$ for a (ZX) ZnO film on a (ZX) silicon substrate.

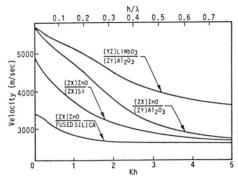

Figure 5.21: (d) SAW properties for various materials.

(d) The phase velocities for four combinations of film and substrate.

The phase velocity of SAW propagating on the layered structure mainly depends on the acoustic properties of the substrate. For a ZnO/glass structure the phase velocity is approximately 3000 m/s at the first peak, $h/\lambda \approx 0.02 - 0.03$. Under a high h/λ value ($h/\lambda \approx 0.4$) the phase velocity is around 2700m/s which is close to that of bulk ZnO. Note that high values of k^2 with high phase velocity are achieved for ZnO single crystal films/R-plane sapphire as shown in Fig.5.22. The maximum value of k^2 is around 5% which is much higher than bulk ZnO and LiNbO$_3$.

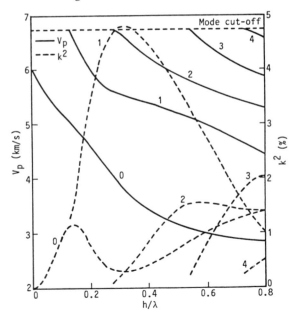

Figure 5.22: Calculated values of V_p and k^2 versus h/λ for $(11\bar{2}0)$ ZnO/$(01\bar{1}2)$Al$_2$O$_3$ structure.

The temperature coefficient of delay time (TCD) for SAW also depends on the property of the glass substrate. Figure 5.23 shows the temperature coefficient of delay for various substrates. It shows that a fused quartz substrate gives a small temperature coefficient at high h/λ; and borosilicate glass, at small h/λ.

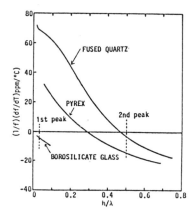

Figure 5.23: Temperature coefficient of delay time in the c-axis oriented ZnO films for various glass substrates.

The propagation loss for SAW on a ZnO/glass structure was measured to be 4 dB/cm at 98 MHz for $h/\lambda = 0.03$. This value is rather large in comparison with single-crystal materials, but considerably smaller than that of ceramic materials. Considering the loss value, polycrystalline ZnO films are regarded as usable for SAW devices operating at frequencies of up to several hundred MHz. Low loss value has been achieved for epitaxial ZnO single crystal films.

Table 5.7 shows the summary of the physical properties for sputtered ZnO thin films.

Thin film ultrasonic transducers.:   C-axis orientated ZnO thin films show good piezoelectric properties. It is therefore possible to construct ultrasonic transducers and composite resonators from these materials (109,110).

Figure 5.24 shows an example of a ZnO thin film sensor (111). The sensor is composed of a ZnO thin film composite resonator with a ZnO thin film/ thin glass substrate. The composite resonator detects the ultrasonic sound generated from the gas flow and senses the flow level of gasses. An integrated oscillator is also made with these thin film resonators.

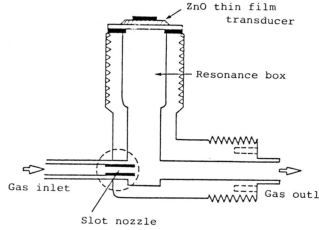

Figure 5.24: Gas flow sensor with ZnO thin film transducer.

Deposition method* (source)	Substrate	Deposition temp. (°C)	Structure**	Film properties***	Ref.
DC-TSP (ZnO)	sapphire	200	PC	$k_t \sim 0.22$ (500MHz) $k_s \sim 0.18$ (500MHz)	a, b
DC-SP (ZnO)	sapphire	75	PC	$k_t \sim 0.25$ (0.4-1.8GHz)	c
RF-SP (ZnO)	FQ		PC	$k^2_{SAW} \sim 0.0025$ (hk=1.1)	d
RF-SP (ZnO	sapphire	200	PC	$k_t = 0.2 \sim 0.25$	e
DC-TSP (ZnO)	FQ	175	PC	$k^2_{SAW} \sim 0.024$	f
PF-SP (ZnO)	FQ	350	PC		g
RF-MSP (ZnO)	glass	100 ∿ 200	PC	$k_t = 0.2 \sim 0.24$ $v_t = 5900 \sim 6900$ m/s $c^* = 8 \sim 9$	h
DC-SP (Zn, Al)	glass	100 ∿ 200	PC	$k_s = 0.05 \sim 0.08$ $v_s = 2600 \sim 3000$ m/s	i
DC-SP (ZnO)	FQ	180 ∿ 260	PC	$k_t = 0.18 \sim 0.25$	j
CVD	sapphire	700 ∿ 750	PC	$k^2_{SAW} \sim 0.0139$	k
RF-MSP (ZnO)	SiO₂	250 ∿ 320	PC		l
RF-SP (ZnO)	sapphire	600	SC	$\sim 2.4 \times 10^3$ Ωcm (0001)sapphire $\mu_H \sim 2.6-28$ cm²/v.s	m
RF-MSP (ZnO, Li)	sapphire	400	SC	$v_{SAW} \sim 5160$ m/s $k^2_{SAW} \sim 0.035$	n
IP (Zn)	glass	50 ∿ 300	PC		o

* SP, diode sputtering; TSP, triode sputtering; HSP, hemispherical sputtering; MSP, magnetron sputtering; IP, ion plating.

** PC, polycrystalline; SC, single crystal.

*** k_t, longitudinal mode coupling; k_s, shear mode coupling; k^2_{SAW}, SAW effective coupling (=2ΔV/V).

Table 5.7: Properties of ZnO thin films.

REFERENCES for Table 5.7.

a. Foster, N.F., in Handbook of Thin Film Technology, (L.I. Maissel and R. Glang, eds.) 15-1, McGraw-Hill, New York (1970).
b. Foster, N.F., J. Appl. Phys., 40: 4202 (1969).
c. Denburg, D.L. IEEE Trans. Sonics and Ultrasonics, Su-18: 31 (1971).
d. Evans, D.R., Lewis, M.F., and Patterson, E., Electron Lett., 7: 557 (1971).
e. Larson, L.D., Winslow, D.K., and Zitelli, L.T., IEEE Trans. Sonics and Ultrasonics, Su-19: 18 (1972).
f. Hickernell, F.S., J. Appl. Phys., 44: 1061 (1973).
g. Hickernell, F.S., J. Solid State Chemistry, 12: 225 (1975).
h. Ohji, K., Tohda, T., Wasa, K., and Hayakawa, S., J. Appl. Phys., 47: 1726 (1976).
i. Wasa, K., Hayakawa, S., and Hada, T., IEEE Trans. on Sonics and Ultrasonics, Su-21: 298 (1974).
j. Chubachi, N., Proc. IEEE, 64: 772 (1976).
k. Tiku, S.K., Lau, C.K., and Lakin, K.M., Appl. Phys. Lett., 33: 406 (1980).
l. Shiozaki, T., Ohnishi, S., Hirokawa, Y., and Kawabata, A., Appl. Phys. Lett., 33: 318 (1978).
m. Mitsuyu, T., Oho, S., and Wasa, K., J. Appl. Phys., 51: 2464 (1980).
n. Mitsuyu, T., Yamazaki, O., Ohji, K., and Wasa, K., J. Crystal Growth, 42: 233 (1982).
o. Machida, K., Shibutani, M., Murayama, Y., and Matsumoto, M., Trans. IEEE 62: 358 (1979).

Another interesting application of the ZnO thin film transducers is an ultrasonic microscope first proposed by Quate. Figure 5.25 shows the construction of the ultrasonic microscope developed by Chubachi (112). The thin film transducer easily generates a focused ultrasonic beam onto the small test sample in a liquid medium.

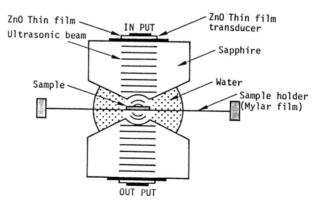

Figure 5.25: Construction of thin film ultrasonic microscope (Kushibiki, Chubacji (1985), (112)).

Surface acoustic wave (SAW) devices: Thin film SAW devices are one of the most interesting thin film electronic components (113). The devices include band-pass filters, resonators, voltage-controlled oscillators, and convolvers in a frequency range of 10 MHz to GHz. The ZnO thin-film SAW video intermediate frequency (VIF) filters for color TV sets and VTR are now widely in production (114,115).

Figure 5.26 shows a typical construction and bandpass characteristics of ZnO thin film SAW VIF filters designed for the NTSC American band. The ZnO thin films are composed of a polycrystalline structure with a c-axis orientation. They are deposited on a borosilicate glass substrate using hemispherical rf-sputtering. The interdigital electrodes are made of evaporated Al thin films. The thickness of the ZnO is 1.5 μm corresponding to 3 percent of the SAW wavelength which induces strong electromechanical coupling in the layered structure. ZnO thin film SAW filters reduce the number of electronic components in color TV sets. The layered structure, ZnO/borosilicate glass, strongly improves the temperature stability of bandpass properties due to the suitable selection of borosilicate glass composition.

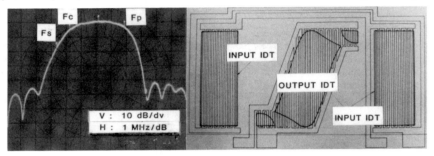

Figure 5.26: Typical configuration and frequency response of ZnO thin film SAW VIF filter for TV sets. $F_p = 45.75$MHz.

Single crystal ZnO thin films epitaxially grown on sapphire are used for making the ZnO thin film SAW devices for the UHF region, since they exhibit small propagating loss of SAW. Such UHF SAW devices are used for various communication systems. The layered structure of ZnO/sapphire yields high electromechanical coupling with high SAW phase velocity. The high phase velocity allows the high frequency operation of the SAW devices. Suitable design of a SAW filter makes it possible to operate at a frequency of 4.4GHz with an IDT finger width of 0.5μm. The frequency response is shown in Fig. 5.27.

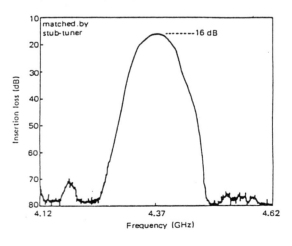

Figure 5.27: Frequency response of SAW filter using single-crystal ZnO thin film on sapphire.

Single crystal ZnO thin films are also excellent wave-guide materials for planar guided light, and are useful for making an acoustooptic Bragg diffractor as shown in Fig. 5.28 (116). The high frequency operation of SAW causes the high diffractive angle. Since the Bragg angle corresponds to the acoustic frequency, this diffractor works as a spectrum analyzer. Figure 5.29 shows typical operation of the diffractor. Each spot is diffracted light by the acoustic wave of respective frequency.

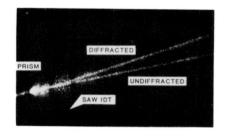

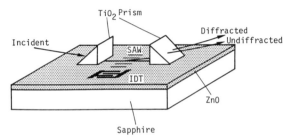

Figure 5.28: Schematic illustration and a typical experiment of the A-O deflector using ZnO thin film on sapphire structure.

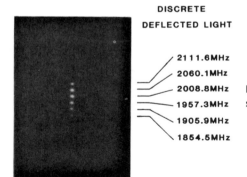

Figure 5.29: Deflected light beam positions as a function of input frequency.

ZnO thin film electronic components: The semiconductive properties of ZnO thin films are used for making ZnO/Si heterojunction photodetectors (117), ZnO/Bi O thin film varistors (118), and ZnO/Pt gas sensors with Schottky structure. Figure 5.30 shows typical properties of the ZnO/Si heterojunction photodetectors. ZnO thin film devices using their semiconducting properties are not widely used as yet. The semiconducting properties are controlled by co-sputtering of an impurity using magnetron sputtering. Recent work done suggests that transparent ZnO thin films of high conducting properties can be produced by magnetron sputtering.

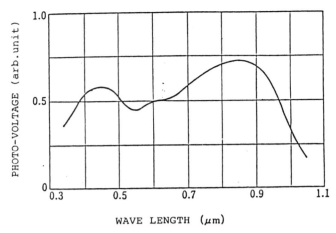

WAVE LENGTH (μm)

Figure 5.30: Typical photoresponse of ZnO thin film/p-Si heterojunction structure.

5.1.2 Sillenite thin films

Crystals of the $\gamma - Bi_2O_3$ family, called sillenites, are attractive materials having strong electro-optical effects, acousto-optical effects, and piezoelectricity (119,120). The sillenites are generally composed of Bi_2O_3 and foreign oxides such as GeO_2, SiO_2, and PbO. $Bi_{12}GeO_{20}$ (BGO), $Bi_{12}SiO_{20}$(BSO), and $Bi_{12}TiO_{20}$ (BTO) are known as the sillenites. Their typical crystal properties are shown in Table 5.8.

Crystal system	23(bcc)
Space group	I23
Lattice constant	a=10.1455 A
Melting point	935°C
Dielectric constant	$\varepsilon_{//}^{S}$ =34.2x10^{-11} F/m
Density	9.2 g/cm^3
Refractive index	2.5476
Optical transparency	0.45 - 7.5 um
Piezoelectric constant	e_{14}=0.99 C/m^2
Acoustic velocity	3.33x10^3 m/s (<111>longitudinal wave)

Table 5.8: Physical properties of $Bi_{12}GeO_{20}$ single crystal.

Thin films of sillenites are deposited by rf-sputtering from the compound target of sillenites. Bi shows the highest vapor pressure in the composition and will exceed 10^{-5} Torr when the substrate temperature is higher than 450°C. this suggests that stoichiometry will be achieved by sputtering from a target of the stoichiometric composition at a substrate temperature below 450°C.

5.1.2.1 Amorphous, Polycrystal Films: Thin films of sillenites with an amorphous and/or polycrystalline structure are made by rf sputtering from a sillenite target. Figure 5.31 shows the construction of the sputtering system. Typical sputtering conditions are shown in Table 5.9 (11). The target is made of a ceramic disk with the stoichometric composition sintered at 800°C for 4 hr in air.

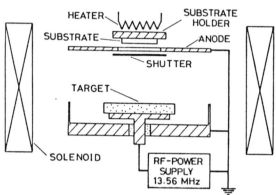

Figure 5.31: Electrode configuration of the sputtering system for the deposition of $Bi_{12}GeO_{20}$ thin films.

Target dimension	30 mm diam.
Target–substrate spacing	25 mm
Sputtering gas	$Ar + O_2$ (1 : 1)
Gas pressure	5×10^{-2} Torr
RF power density	~1 W/cm^2
Magnetic field	100 G
Substrate temperature	350–550°C
Growth rate	~0.5 $\mu m/h$
Film thickness	1.5–4 μm

Table 5.9: Sputtering conditions.

The crystallographic structure of the sputtered films strongly depends on the substrate temperature during deposition. In the case of BGO films, sputtered films show an amorphous phase at a substrate temperature below 150°C. At the substrate temperature of 150 to 350°C, the sputtered films show a polycrystalline form of a metastable δ-phase (121). The $\delta -$ phase shows a face centered cubic structure which does not exhibit piezoelectricity. At substrate temperatures above 400°C, the γ phase appears. Between 350 to 400°C mixed phases of δ and γ phase appear. Typical reflection electron diffraction patterns and electron micrographs of sputtered BGO films are shown in Fig. 5.32.

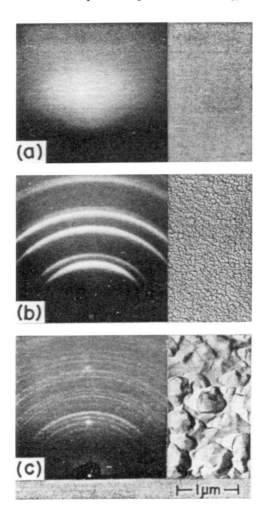

Figure 5.32: Typical reflection electron diffraction patterns and electron micrographs of $Bi_{12}GeO_{20}$ films sputtered onto glass substrates: (a) film sputtered at about 100°C with growth rate of 0.25μm/hr, amorphous state; (b) film sputtered at 200°C, 0.20μm/hr, fcc form; (c) film sputtered at 400°C, 0.15μm/hr, bcc form.

The crystallographic properties of other sputtered sillenite films are generally similar to those of BGO films. The sputtered BLO films, however, show preferential orientation as seen in Fig. 5.33 (15). The direction of the preferred orientation is (310) axis, perpendicular to the substrate.

These sputtered films are semitransparent with a yellow to light brown color. Figure 5.34 shows a typical optical absorption spectra of the sputtered BGO films (11). The optical absorption edge for the amorphous films is about 510μm, the δ phase 620μm and the γ phase 620μm. The absorption edge for the sputtered phase films is nearly equal to the bulk single crystal value 450μm (14). In the infrared region a broad absorption is observed at 20μm as shown in Fig. 5.35. For sputtered phase films several weak absorptions are superposed on the broad absorption spectrum which relates to the characterized lattice vibration of γ -phase sillenites.

Figure 5.33: Typical electron diffraction pattern of $Bi_{12}PbO_{19}$ sputtered onto glass substrate.

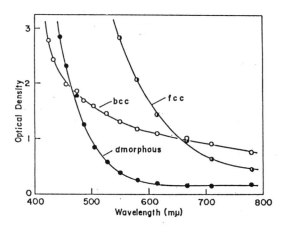

Figure 5.34: Optical absorption spectra in the visible region for three types of sputtered $Bi_{12}GeO_{20}$ films.

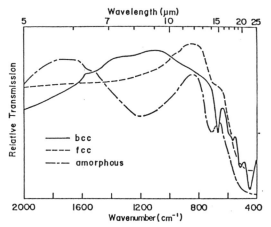

Figure 5.35: Infrared transmission spectra for three types of sputtered $Bi_{12}GeO_{20}$ films.

5.1.2.2   Single Crystal Films: Single crystal films of $\gamma - Bi_2O_3$ compounds are epitaxially grown on single crystal substrates by rf-diode sputtering (14). One example is the BTO thin films on BGO single crystal substrates. Table 5.10 shows some crystal properties of BTO and BGO. The structure BTO/BGO provides an optical waveguide for optical integrated circuits since the refractive index of BTO is greater than that of BGO. The small lattice mismatch ($\simeq 0.3\%$) between BTO and BGO leads to the growth of excellent quality, single crystal BTO on BGO substrates. These substrates are sliced from a single crystal parallel to the (110) plane and then polished by $0.3\mu m$ alumina powder. The epitaxial growth is carried out by the rf diode sputtering system. Since the epitaxial temperature of $\gamma - Bi_2O_3$ compound thin films is in the range of 400 to 500°C, the reevaporation of Bi from the sputtered films must be considered to have stoichiometric films. Table 5.11 shows the Bi/Ti atomic ratios for films sputtered at a substrate temperature of 400 and 500°C from three different targets. It shows the compositions of the films sputtered at 400°C are almost equal to that of each target. Films sputtered at 500°C show a remarkable decrease of the Bi/Ti atomic ratio due to the expected reevaporation of Bi. Their crystalline structures for various substrate temperatures are summarized in Fig. 5.36. Single crystalline γ phase films are epitaxially grown from the stoichiometric target at 400 to 450°C. From the Bi-rich target $(9Bi_2O_3.TiO_2)\gamma$ -phase films are obtained at above 450°C From the Ti-rich target $(4Bi_2O_3.TiO_2)$ mixed phases of the γ phase and $Bi_4Ti_3O_4$ appear. Similar mixed phases are also observed in sputtered films from the stoichiometric target at a substrate temperature above 500°C.

Single crystal films obtained from the stoichiometric target are transparent with a smooth surface as shown in Fig. 5.37. The films act as wave guides. Figure 5.38 shows a guided beam of He-Ne laser light (6328Å) which is fed into the film by a rutile prism. The film thickness is $2.4\mu m$ and the waveguide mode is TE_o.

	Crystal structure	Lattice constant (Å)	Melting point (°C)	Refractive index
$Bi_{12}TiO_{20}$	bcc, space group I23	10.176	~930	2.5619
$Bi_{12}Geo_{20}$	bcc, space group I23	10.1455	~935	2.5476

Table 5.10: Properties of $Bi_{12}TiO_{20}$ and $Bi_{12}GeO_{20}$.

Substrate tempera- ture (°C)	Target composition		
	$9Bi_2O_3 \cdot TiO_2$ (Bi/Ti = 18)	$6Bi_2O_3 \cdot TiO_2$ (Bi/Ti = 12)	$4Bi_2O_3 \cdot TiO_2$ (Bi/Ti = 8)
400	17.3	12.4	7.1
500	13.3	6.2	5.3

Table 5.11: Bi/Ti atomic ratios of sputtered films determined by EPMA.

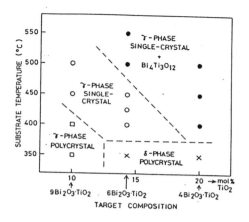

Figure 5.36: Crystalline structures of sputtered films.

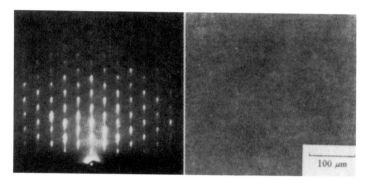

Figure 5.37: Typical electron diffraction pattern and photo-micrograph of epitaxially grown $Bi_{12}TiO_{20}$ onto $Bi_{12}GeO_{20}$ substrate.

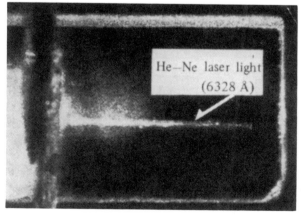

Figure 5.38: Guided beam of He-Ne laser light on the $Bi_{12}TiO_{20}/Bi_{12}GeO_{20}$ structure.

5.1.3 Perovskite Dielectric Thin Films

The perovskite structure observed in ABO_3 type compounds such as $BaTiO_3$ has ferroelectricity similar to the $\gamma - Bi_2O_3$ family (122). Figure 5.39 shows a typical structure of the perovskite crystal structure. Thin films of the perovskites including $BaTiO_3$, $PbTiO_3$, PZT ($PbTiO_3 - PbZrO_3$), and PLZT [(Pb,La)(Zr,Ti)O_3] have been studied in relation to making thin films of dielectrics, pyroelectrics, piezoelectrics, and electro-optic materials.

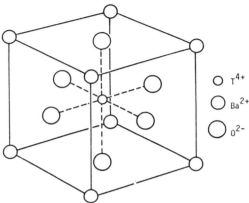

Figure 5.39: Crystal structure of perovskites, $BaTiO_3$.

Thin films of perovskites with a polycrystalline or single crystal structure are deposited by rf-sputtering from a sintered ceramic target. Substrates for deposition of polycrystalline films are Pt or fused quartz. Single crystals of MgO, $SrTiO_3$, sapphire, and spinel are used for epitaxial growth of single crystal perovskite films. The growth temperature of the perovskite structure is 600 to 700°C, thus the substrate temperature must be higher than 600°C during the deposition. However, sputterings are sometimes done at a low substrate temperature of 300°C and the resultant films are post- annealed in air at 600 to 750°C in order to have the perovskite structure of polycrystalline form.

5.1.3.1 $PbTiO_3$ Thin Films:

Deposition: Thin films of perovskite materials including $PbTiO_3$, PZT, and PLZT have been prepared by sputtering and/or by chemical vapor deposition.

Bickley and Campbell have previously deposited mixed films of PbO and TiO_2 by reactive sputtering from a composite lead titanium cathode in an oxidizing atmosphere. They used a conventional dc diode sputtering system, and the mean permitivity of the resultant films was 33 when chemical composition was $PbTiO_3$. This value, however, is much smaller than that of true $PbTiO_3$ compound. The as-grown films gave little information about the formation of $PbTiO_3$. In order to synthesize $PbTiO_3$, the substrate temperature must normally be more than 600°C (123).

The structure and dielectric properties of these sputtered films may vary with the sputtering system. The growth of $PbTiO_3$ will be achieved, even at low substrate temperatures of 200°C or less by magnetron sputtering deposition under low working pressure (124). Typical sputtering conditions are shown in Table 5.12.

Sputtering system	dc-magnetron
Target	Pb-Ti composite
Sputter gas	$Ar/O_2 = 1$ $(6{\times}10^{-4}$ Torr$)$
Substrate	glass
Substrate temperature	150 - 300 °C
Deposition rate	30 - 600 Å/min

Table 5.12: Typical sputtering conditions for the deposition of $PbTiO_3$ thin films by DC magnetron system.

As shown in Table 5.12, the composite lead-titanium cathode was used for deposition. During sputtering, PbO and TiO_2 are codeposited onto the substrate and mixed films of PbO and TiO_2 are fabricated. The chemical composition of the sputtered films is controlled by the ratio of Pb/Ti in the composite lead-titanium cathode. Figure 5.40 shows the dielectric properties of the mixed films for various chemical compositions. It shows that the permitivity maximum observed in the chemical composition of $PbTiO_3$ is higher than the permitivity of PbO or TiO_2. The temperature variation of the permitivity also shows the maximum at about 490°C, as indicated in Figure 5.41, which is the value expected for $PbTiO_3$. These electrical properties suggest that $PbTiO_3$ is synthesized in mixed films prepared by magnetron sputtering (123).

PbO and TiO_2 phases were detected by electron diffraction analysis of the mixed films. Thus the remainder could be amorphous $PbTiO_3$. The content X_{PT} of $PbTiO_3$ can be estimated from

$$X_{PT} = \frac{\log(\varepsilon_M/X_{TiO_2}X_{PbO})}{\log(\varepsilon_{PT}/X_{TiO_2}X_{PbO})}, \quad (5.4)$$

if we assume that Lichteneker's empirical logarithmic mixing rule is established between ε_{PbO}, ε_{TiO_2}, and ε_{PT}, where these values are the permittivities of PbO, TiO_2, $PbTiO_3$ and the sputtered mixed films, respectively and X_{PbO}, X_{TiO_2} and X_{PT} are the proportions by volume of PbO, TiO_2 and $PbTiO_3$ respectively, so that $X_{PbO} + X_{TiO_2} + X_{PT} = 1$. Putting $\varepsilon_M \simeq 120$, $\varepsilon_{PbO} \simeq 25$ and $\varepsilon_{TiO_2} \simeq 60$, we have $X_{PT} \simeq 0.7$. This estimate suggests that 70% of the sputtered film is $PbTiO_3$.

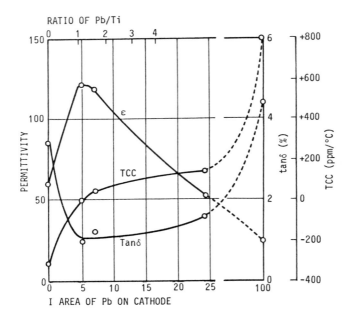

Figure 5.40: The dielectric properties of Pb-Ti-O films 3000Å thick on 7059 glass deposited by a magnetron sputtering system (measured at 1 MHz and room temperature).

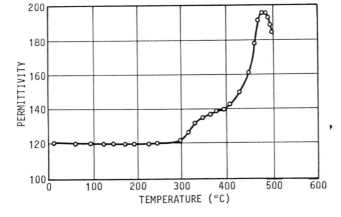

Figure 5.41: The temperature variation of the permitivity of Pb-Ti-O films 3000Å thick on 7059 glass deposited by a magnetron sputtering system (measured at 1 MHz).

In magnetron sputtering, the oxides of the cathode metals, i.e. PbO and TiO_2, are initially formed at the cathode surface. These oxides will be co-sputtered, and some fraction may be sputtered as molecules of PbO and TiO_2 which are deposited on the substrates. These sputtered molecules have an energy of 1 - 10 eV when they strike the substrates since they suffer few collisions with gas molecules in transit and have approximately the same energy as when they were removed from the cathode surface. The activation energy in the chemical reaction $PbO + TiO_2 \rightarrow PbTiO_3$ is on the order of 1 eV. Thus when these sputtered PbO and TiO_2 molecules collide with each other on the substrates, PbTiO$_3$ can conceivably be synthesized even at low substrate temperatures. In conventional sputtering (i.e., higher pressure rf diode) the reaction between PbO and TiO$_2$ might not be possible at low substrate temperature because the energy of the sputtered PbO and TiO$_2$ molecules is greatly reduced on the substrate due to collisions between the sputtered molecules and gas molecules in transit.

The rf-magnetron sputtering system is also employed for the preparation of the PbTiO$_3$ thin films (125). Typical sputtering conditions are shown in Tab. 5.13. A powder of PbTiO$_3$, which is put in a stainless steel dish, is used as the target. X-ray diffraction analyses suggest that the PbTiO$_3$ thin films deposited on a cooled glass substrate (liq. N$_2$≃ room temperature) are amorphous structures (a − PbTiO$_3$) with Pb crystallites. Pb crystallites make the film electrically conductive [$\sigma > 10(\Omega cm)^{-1}$]. They disappear upon annealing the film above 220°C. The formation of Pb crystallites may be attributed to extremely rapid quenching of adsorbed atoms on the substrate in the sputter deposition process. The surface migration of the adsorbed atoms is not activated on the cooled substrate. The adsorbed atoms are condensed so rapidly that Pb crystallites (whose free energy was closer to that of the vapor) are formed.

Sputtering system	rf-magnetron
Target	PbTiO$_3$ compounds
Sputter gas	Ar/O$_2$ = 1 (5×10^{-3} Torr)
Substrate	glass, sapphire, MgO, SrTiO$_3$
Substrate temperature	liq. N$_2$ - 700 °C
Deposition rate	50 - 70 Å/min

Table 5.13: Typical sputtering conditions for the deposition of PbTiO$_3$ thin films by rf-magnetron sputtering.

Heating the substrate activates the surface migration of the adsorbed atoms on the substrate. The PbTiO$_3$ deposited on the heated substrates (> 200°C) did not contain Pb crystallites. The X-ray diffraction pattern of a − PbTiO$_3$ deposited at 200°C is shown in Fig. 5.42. The diffraction pattern is a halo and suggests that the film exhibits a uniform amorphous configuration (which contained no crystallites). Activating the surface migration can cause the disappearance of Pb crystallites and the formation of uniform amorphous configurations.

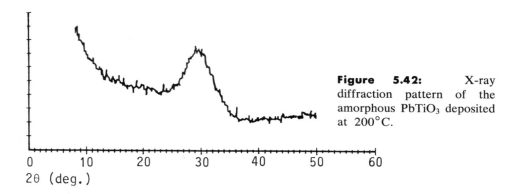

Figure   5.42:   X-ray diffraction pattern of the amorphous PbTiO₃ deposited at 200°C.

Further activating of the surface migration may cause the formation of crystalline configurations. When the substrate is heated above 200°C during the deposition, the film is partially crystallized. Figure 5.43 shows the X-ray diffraction pattern of the film deposited at 500°C. The observed peaks can be indexed by pyrochlore $Pb_2Ti_2O_6$ (ASTM card 26- 142). The films deposited above 500°C are a mixed polycrystalline of pyrochlore and perovskite. Figure 5.44 shows the schematic phase diagram of the sputter-deposited $PbTiO_3$ film.

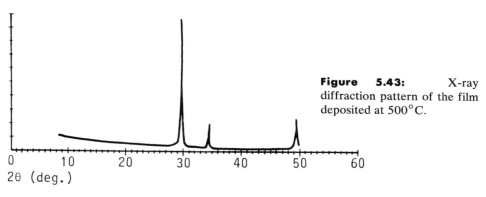

Figure   5.43:   X-ray diffraction pattern of the film deposited at 500°C.

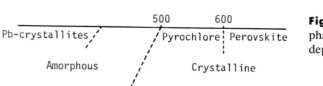

Figure   5.44:   Schematic phase diagram of the sputter deposited $PbTiO_3$.

Figure 5.45 shows the X-ray diffraction pattern of the a — PbTiO₃ film annealed at 600°C. All of the X-ray diffraction peaks can be indexed by perovskite PbTiO₃ (ASTM card 6-0452). No diffraction peak attributed to pyrochlore is observed. The a — PbTiO₃ film changes to a polycrystalline perovskite when it is annealed above 520°C whereas the film deposited at 600°C is a mixed polycrystalline of pyrochlore and perovskite. The pure perovskite PbTiO₃ film is more easily prepared by annealing the a — PbTiO₃ film.

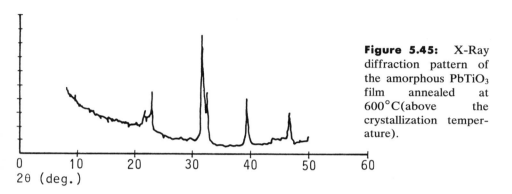

Figure 5.45: X-Ray diffraction pattern of the amorphous PbTiO₃ film annealed at 600°C(above the crystallization temperature).

Figure 5.46 shows the X-ray diffraction pattern of the a — PbTiO₃ film annealed at 480°C (below the crystallization temperature). This film is partially crystallized. The diffraction peaks can be indexed by a mixture of perovskite and pyrochlore. Two kinds of crystallites grow below the crystallization temperature, whereas only the perovskite crystallites grow above the crystallization temperature. We may consider that the sputter deposited a — PbTiO₃ contains crystallites of perovskite-like and/or pyrochlore-like microstructures.

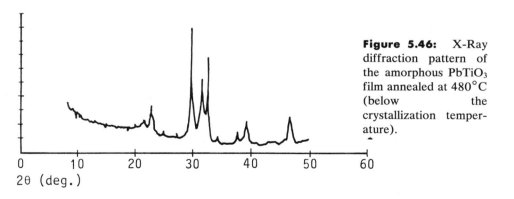

Figure 5.46: X-Ray diffraction pattern of the amorphous PbTiO₃ film annealed at 480°C (below the crystallization temperature).

Polycrystalline or single crystal thin films of PbTiO₃ are deposited at substrate temperatures higher than the crystalline temperature. Figure 5.47 shows a typical X-ray reflection spectra obtained from the target powder and the epitaxial PbTiO₃ thin films on a c-axis sapphire sputtered at 620°C. The epitaxial relationship is

$$(111)PbTiO_3 \parallel (0001)sapphire. \qquad (5.5)$$

A typical RED pattern and optical transmission spectrum of the PbTiO₃ thin films epitaxially grown on sapphire are shown in Figs. 5.48 and 5.49, respectively. The (100) PbTiO₃ thin films will be epitaxially grown on the (100) surface of MgO and/or SrTiO₃ single crystal substrates.

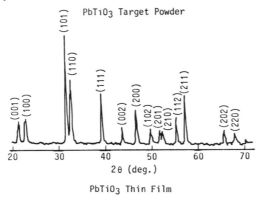

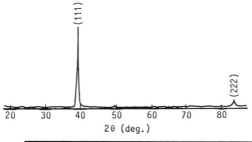

Figure 5.47: Typical X-ray reflection spectra from PbTiO₃ target powder and epitaxial PbTiO₃ single crystal films on (0001) sapphire.

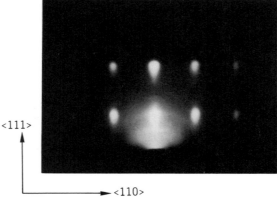

Figure 5.48: A typical electron diffraction pattern of sputtered PbTiO₃ thin film epitaxially grown on (0001) sapphire.

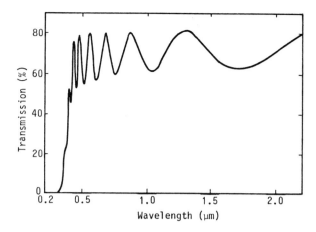

Figure 5.49: Optical transmission spectrum of epitaxial PbTiO$_3$ thin films about 0.4μm thick.

Polycrystalline films with a perovskite structure are grown at a substrate temperature of 450 to 600°C. The polycrystalline films show a preferred orientation of (110) which corresponds to the densest packed plane.

Since the vapor pressure of Pb becomes high at the epitaxial temperature of about 600°C, the resultant PbTiO$_3$ thin films often show a deficiency of Pb. In order to keep the stoichiometric composition in the sputtered PbTiO$_3$ thin films, a multi-target sputtering system is used for the deposition (126). The schematic illustration of the multi-target sputtering system is shown in Fig. 5.50.

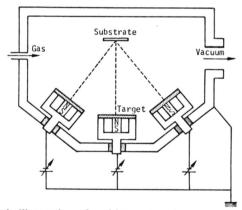

Figure 5.50: Schematic illustration of multi-target system.

Separate magnetron cathodes are equipped and metal targets of each component are placed on them. The targets then focus on the substrate. The normal line of the substrate makes an angle of 30 degrees to that of each target. The sputtering rate of each target is individually controlled by a dc power supply. Reactive sputtering is carried out by introducing a mixed gas of oxygen and argon. The details of the multi-target sputtering apparatus and sputtering conditions are summarized in Table 5.14.

Target	Pb, Ti metal
Target diameter	60 mm
Target-substrate spacing	100 mm
Sputtering gas	Ar/O_2 = 1/0 - 1/1
Gas pressure	0.5 - 10 Pa
Input power	Pb: 0 - 15 W
	Ti: 0 - 200 W

Table 5.14: Sputtering conditions for a multi-target system.

In general, the sputtering rate of the metal target decreases by increasing the oxygen partial pressure. When the oxygen partial pressure is high, the surface of the target is oxidized and the sputtering rate decreases remarkably. The effect of the oxygen partial pressure on the sputtering rate for Pb and Ti targets is shown in Fig. 5.51. The dependence of the sputtering rate upon oxygen gas pressure differs greatly between the Pb and Ti targets. The sputtering rate decreases more rapidly at the Ti target than at the Pb target. The deposition rate of lead oxide is much larger than that of titanium oxide. In order to provide the same deposition rate for Ti and Pb in a fully oxidizing atmosphere, the sputtering conditions are fixed at an oxygen partial gas pressure of 1 Pa, input power of $5\approx15W$ for the Pb target and $100\approx200W$ for the Ti target, respectively.

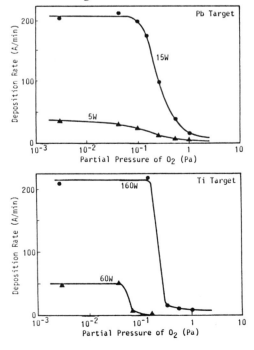

Figure 5.51: The relation between the oxygen partial pressure and the deposition rate for Pb and Ti target.

Figure 5.52 shows the composition and crystallinity of the films obtained at various substrate temperatures and incident Pb/Ti ratios. The composition and crystallinity of the films were examined by EPMA and X-ray diffraction methods, respectively. The compositional Pb/Ti ratios of thin films grown at 50°C can be regarded as incident Pb/Ti ratios to the substrate, since reevaporation is negligible. The epitaxial growth of the perovskite structure is found at high substrate temperatures. When the substrate temperature is 700°C, excess Pb (greater than Pb/Ti≈1.1) seemed to reevaporate from the substrate. In the event that only the Pb target is sputtered, film growth was not observed at 700°. This indicates that the affinity between Pb and Ti prevents the evaporation of Pb as seen in the sintering of $PbTiO_3$ ceramics. The crystallinity of the epitaxial thin films is evaluated by X-ray diffraction peak intensity. The index of the diffracted X-ray used is (111) and the intensity is normalized by that of sapphire (0006). Figure 5.53 shows the X-ray peak intensity of the films as a function of the incident Pb/Ti ratio. The crystallinity of the films seems to improve as the incident Pb/Ti ratio approaches about 1.1. Though the compositional Pb/Ti ratio of the films at the incident Pb/Ti ratios of 2.3, 1.6 and 1.2 are nearly equal at a substrate temperature of 700°C the X-ray peak intensities of these three films are obviously different. An optimum incident ratio of Pb to other components does exist for preparing epitaxial thin films of good quality.

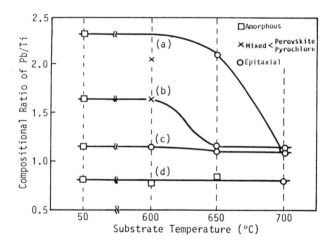

Figure 5.52: Composition and crystallinity of the films obtained at various substrate temperature and incident Pb/Ti ratio. The incident Pb/Ti ratio is (a) 2.3, (b) 1.6, (c) 1.2 and (d) 0.8.

Electrical properties: Figure 5.54 shows the temperature dependence of the dielectric constant of as-deposited $a - PbTiO_3$ film deposited at 300°C. Two anomalies are observed at 520 and 480°C. The crystallization temperature is 520°C and one of the anomalies is attributed to it. Figure 5.55 shows the temperature dependence of the dielectric constant of the annealed (above 520°C) and crystallized film. An anomaly, which is caused by the phase transition of ferroelectric perovskite $PbTiO_3$, is observed at 480°C. It is considered that the dielectric anomaly (at 480°C) of the as-deposited $a - PbTiO_3$ is attributed to the phase transition of ferroelectric perovskite $PbTiO_3$. This dielectric anomaly of the as-deposited $PbTiO_3$ is larger and sharper than that of the crystallized one. The as-deposited amorphous film partially crystallizes and contains perovskite PbTiO₃ crystallites.

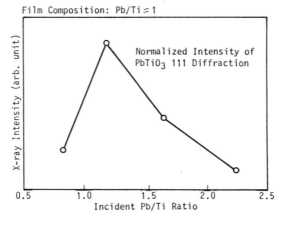

Figure 5.53: The relation between the incident Pb/Ti ratio and X-ray diffraction peak intensity of the epitaxial films prepared at substrate temperatures of 650°C and 700°C.

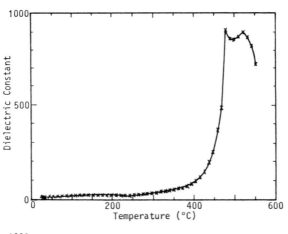

Figure 5.54: Temperature dependence of the dielectric constant of the amorphous PbTiO₃ film deposited at 300°C. The film thickness was about 1.5μm. The measuring ac (10kHz) voltage is 0.1 V rms.

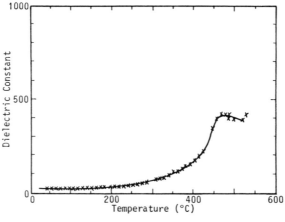

Figure 5.55: Temperature dependence of the dielectric constant of the PbTiO₃ crystallized film (measured at 10 kHz).

It is reported that the roller quenched a – $PbTiO_3$ and its annealed platelets are in a "heavily pressed state" and the dielectric anomaly, which is caused by the phase transition of ferroelectric perovskite $PbTiO_3$, shifted to a lower temperature. The sputter deposited a – $PbTiO_3$ and its annealed film exhibit this anomaly at the same temperature and no temperature shift is observed. Therefore the the perovskite $PbTiO_3$ crystallites, which grow in the sputter deposited a – $PbTiO_3$, are considered to be in a "stress-free state". This property is the most significant difference between sputtering-deposited and roller-quenched a – $PbTiO_3$. Scanning electron microscopy suggests that sputter deposited a – $PbTiO_3$ contains many voids. These voids may compensate for the stress. This stress-free growth of the perovskite $PbTiO_3$ crystallites enables us to study pure "grain-size effect" of ferroelectrics.

The dielectric anomaly in $PbTiO_3$ thin films is clearly observed in epitaxially grown films. A typical result is shown in Fig. 5.56 (127). This anomaly is observed at 490°C which corresponds to the anomaly temperature for bulk $PbTiO_3$.

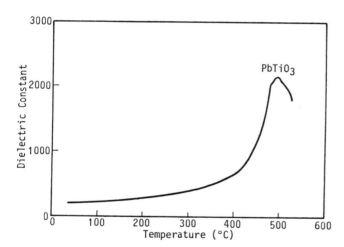

Figure 5.56: Temperature dependence of the dielectric constant of $PbTiO_3$ thin film about 0.4μm thick epitaxially grown on (0001) sapphire (measured at 100 kHz).

The piezoelectric properties of $PbTiO_3$ thin films have been studied in detail by Kushida et al. for c-axis oriented films (128). Measurements of the electromechanical coupling factor k_t show c-axis oriented $PbTiO_3$ films were formed on patterned Pt electrode films embedded in the $SrTiO_3$ single crystal seeded lateral overgrowth. The structure of the sample is shown in Fig. 5.57. The substrate is a (100) $SrTiO_3$ single crystal plate. The impedance characteristics of the $Au/PbTiO_3/Pt/SrTiO_3$ structure are evaluated by the Mason equivalent circuit shown in Fig. 5.58.

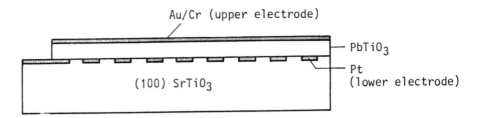

Figure 5.57: Structure of the sample for the measurements of piezoelectric coupling factor for the thickness vibration k_t (Kushida (1987) (128)).

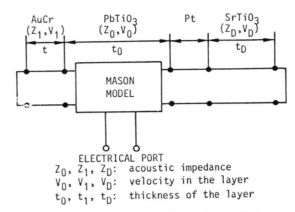

ELECTRICAL PORT
Z_0, Z_1, Z_D: acoustic impedance
V_0, V_1, V_D: velocity in the layer
t_0, t_1, t_D: thickness of the layer

Figure 5.58: Equivalent circuit model used in analysis of the composite resonator (Kushida (1987) (128)).

In the Mason circuit, the electromechanical coupling factor k_t is deduced from the following impedance formula:

$$S = \frac{1}{j\omega C_o}[1 + \frac{k_1^2}{\Theta_0}[\frac{2(1 - \cos \theta_0) + (z_1 + z_D) \sin \Theta_0}{(1 + z_1 z_D) \sin \Theta_o - (z_1 + z_D) \cos \Theta_0}]], \quad (5.6)$$

where C_0 is the constant strain capacitance of the piezo layer, $k_1^2 = h_{33}^2 \varepsilon_{33}/c_{33}$ is the electromechanical coupling constant of the $PbTiO_3$ film (h is the piezoelectric constant, ε the dielectric constant, and c the elastic stiffness), $\Theta_0 = 2\pi f t_0/u_0$ is the acoustic phase in the piezo film, $\Theta_1 = 2\pi f t_1/u_1$ is the acoustic phase in the electrode, $\Theta_D = 2\pi f t_D/u_D$ is the acoustic phase in the substrate, $z_1 = Z_1 \tan \Theta_1/z_0$, $z_D = Z_D \tan \Theta_D/z_0$ and Z_0, Z_1, and Z_D are the acoustic impedances.

The values k_t are evaluated from impedance measurements are shown in Fig. 5.59. The impedance characteristics show the resonant properties of a composite bulk wave resonator. The impedance measurements suggest that the c-axis oriented $PbTiO_3$ thin films $1 - 2\mu m$ thick exhibit a $k_t \approx 0.8$ at a frequency below 350 MHz. This value is extremely large for piezoelectric thin films and is comparable to the value obtained for a PbTiO₃ single crystal (128).

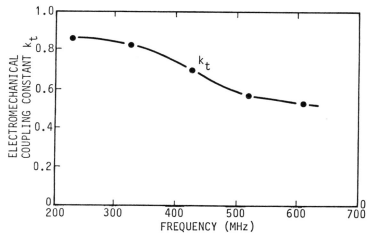

Figure 5.59: Frequency variation of the calculated electro-mechanical coupling constant k_t (Kushida (1987) (128)).

Thin films of PbTiO₃ are also known the pyroelectric materials (126). Figure 5.60 shows the construction of a thin film pyroelectric sensor (129). A small heat capacity with high pyroelectricity increases the pyroelectric current and decreases Johnson noise, and makes for an excellent pyroelectric sensor. Table 5.15 shows the summary of the electrical properties of PbTiO₃ thin films.

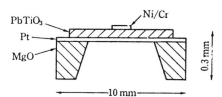

Figure 5.60: Construction of PbTiO₃ thin film pyroelectric sensor for IR detection.

5.1.3.2 PLZT Thin Films:

Deposition: Figure 5.61 shows the phase diagram of PLZT (x/y/x), $[(Pb_{1-x}, La_x)(Zr_y Ti_{1-y})_{1-x/4}O_3]$ ceramics (81). Thin films of PLZT are prepared by the sputter deposition similar to deposition of PbTiO₃ thin films.

Deposition* method	Substrate	Deposition temp. (°C)	Structure**	Film properties	Ref.
DC-MSP	glass	200	a	ϵ^* - 120 (RT) T_c - 490 °C	79
RF-MSP	glass	liq. N_2	a	ϵ^* - 800 (200 °C)	80
RF-MSP	sapphire	580	pc	ϵ^* - 370 (RT) T_c - 490 °C	34
RF-MSP	P_t	630	pc	ϵ^* - 110 (RT) k_t - 0.8	83
RF-MSP	P_t	575	PC	ϵ^* - 97 (RT) pyroelectric coefficient γ - 3×10^{-8} $C/cm^2 K$	84

** a, amorphous;   pc, polycrystalline

* MSP, magnetron sputtering

Table 5.15: Properties of $PbTiO_3$ thin films .

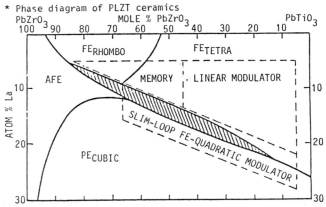

Figure 5.61: Phase diagram of PLZT ceramics (G.H. Haertling and C.E. Land, 1971).

Since the vapor pressure of Pb becomes high at the epitaxial temperature the resultant films often show a deficiency of lead components. Achieving stoichiometry in epitaxial films is much more important for the deposition of complex compounds PLZT. Figure 5.62 shows the typical spectra of XMA obtained from PLZT (9/65/35) thin films and the target. The composition of the sputtered PLZT films is roughly estimated from the spectra. The content of Pb decreases with the increase of substrate temperature as described before. In order to compensate for the lead deficiency, excess PbO, 5 to 10 mole %, should be added to the target. Also note that the degree of Pb deficiency will strongly depend on the target conditions and type of sputtering system. Typical results are shown in Fig. 5.63. Magnetron sputtering with a powder target shows the smallest Pb deficiency.

The ratio of Zr/Ti is close to the target in magnetron sputtering as indicated in Table 5.16. The magnetron discharge permits a lowering of the sputtering gas pressure which helps to lower the epitaxial temperature. This enables one to keep the stoichiometric composition in epitaxial films. Typical sputtering conditions for PLZT thin films are shown in Table 5.17. Sapphire is used for the substrate. With respect to crystal orientation, the (0001) plane, the c-plane of sapphire, is suitable for epitaxial growth when one considers the atomic configuration. The plane has a normal 3-fold axis to the plane and is the same symmetry as the (111) plane of PLZT. Their atomic configurations are shown in Fig. 5.64. The average distances of oxygen atoms are 2.75Å for the sapphire and 2.8Å for the PLZT. The lattice mismatch is about 2%. Their epitaxial relationship is as follows:

$$(111)\text{PLZT} \parallel (0001) \text{ sapphire} \tag{5.7}$$

$$[1\bar{1}0]\text{PLZT} \parallel [10\bar{1}0] \text{ sapphire}$$

The crystal orientation of the epitaxial PLZT thin film on the sapphire substrate is schematically shown in Fig. 5.65.

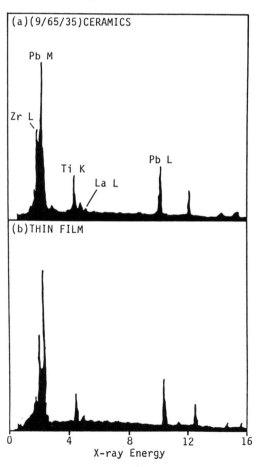

Figure 5.62: XMA patterns of PLZT (9/65/35) ceramic target (a), and sputtered thin films (b).

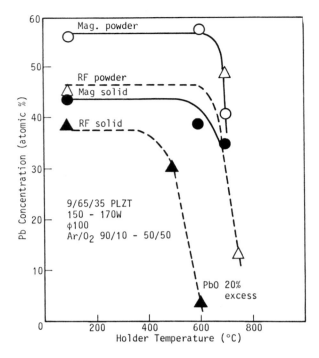

Figure 5.63: Pb concentration vs substrate holder temperature in the compound thin films sputtered from PLZT (9/65/35) target.

Sputtering method	Target	Zr/Ti ratio
Rf-diode	powder	73/27
	plate	73/27
Rf-magnetron	powder	64/36
	plate	64/36

* Target Zr/Ti = 65/35, Substrate temp. 700 °C

Table 5.16: The atomic ratio Zr/Ti in sputtered PLZT (9/65/35) thin films for various sputtering conditions.

Target	PLZT powder
Target diameter	100 mm
Substrate	Sapphire (0001)
Target-substrate spacing	35 mm
Sputtering gas	Ar (60%) + O_2 (40%)
Gas pressure	0.5 Pa
Substrate temperature	500 - 700 °C
Rf power	150 - 250 W
Deposition rate	60 - 100 Å/min

Table 5.17: Typical sputtering conditions for the deposition of PLZT thin films by rf magnetron sputtering.

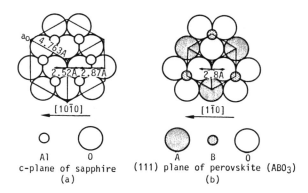

a_0 4.763Å

2.52Å 2.87Å

[10$\bar{1}$0]

2.8Å

[1$\bar{1}$0]

○ Al   ○ O
c-plane of sapphire
(a)

● A   ◉ B   ○ O
(111) plane of perovskite (ABO$_3$)
(b)

Figure 5.64: Planar atomic arrangements of c-plane of sapphire and the (111) plane of cubic perovskite (ABO$_3$).

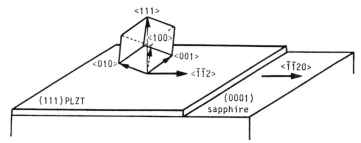

<111>

<100>

<010>   <001>

<$\bar{1}\bar{1}$2>   <$\bar{1}\bar{1}$20>

(111)PLZT   (0001) sapphire

Figure 5.65: Crystal orientation of the epitaxial (111) PLZT film on (0001) sapphire.

Figure 5.66 indicates the crystalline structures of films deposited at various conditions. It shows that the film structure depends primarily on the substrate temperature and is only slightly affected by the growth rate. At substrate temperatures lower than 550°C, a metastable pyrochlore structure appears. The general formula of the pyrochlore is $A_2B_2O_7$, and their films have an intense yellow color. The figure also shows that epitaxial films with a perovskite structure are obtained at substrate temperatures higher than about 550°C. Epitaxial perovskite films are colorless. Figure 5.67 shows a typical RHEED pattern of the epitaxial PLZT film.

The composition of deposited films also depends on the substrate temperature and is independent of growth rate. The solid line (a) in Fig. 5.68 shows the compositional Pb/Ti ratio as a function of the substrate temperature at the growth rate of 80Å/min. The epitaxial perovskite films obtained at above 550°C are almost stoichiometric, while the pyrochlore films obtained at below 550°C are remarkably Pb rich. The excess Pb content in the film is considered to prevent the epitaxial growth of the perovskite structure. These considerations are experimentally confirmed as shown by the dash line (b) in Fig. 5.68. The minimum substrate temperature for epitaxial growth is decreased to 450°C. It is observed that the composition of the PLZT films is close to stoichiometric when a target with less Pb is used. Thus, an epitaxial film growth can be observed, even at lower substrate temperature. This result seems to disagree with Ishida's work using conventional diode sputtering where a Pb-rich target was required. In the magnetron sputtering system, however, the discharge plasma is located primarily near the target, so that Pb in the target is considered to have evaporated as compared to conventional diode sputtering.

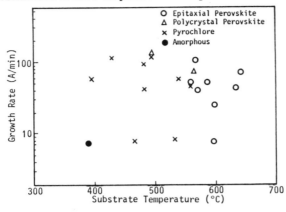

Figure 5.66: Crystalline structures of films deposited on sapphire at various substrate temperature and growth rate.

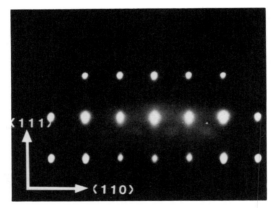

Figure 5.67: RHEED pattern of the epitaxial PLZT (28/0/100) thin film. The thickness is 0.4μm.

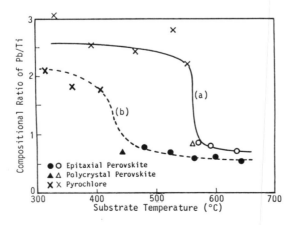

Figure 5.68: Compositional Pb/Ti ratio and crystalline structure of films deposited at various substrate temperatures using PLZT (28/0/100) target (a) and Pb-reduced PLZT (28/0/100) target (b). The growth rate is about 80 Å/min.

In order to keep the correct stoichiometric composition in the sputtered PLZT thin films, the multi-target sputtering system described in Fig. 5.50 is much more useful than the conventional rf-magnetron sputtering system (130). It is also interesting that an artificial superlattice structure composed of multi-layers of different ferroelectric materials can be made by the multi-target sputtering system. Figure 5.69 shows the construction of the multi-target sputtering system for the deposition of the ferroelectric superlattice, PLT-PT structure. Targets of each element are separately positioned facing the substrate. Reactive co-sputtering is carried out by introducing a mixed gas of argon and oxygen. The dc power supply for each target is controlled by a desk-top computer (HP-9835A) and the input power of each target was varied periodically. Typical sputtering conditions are shown in Table 5.18.

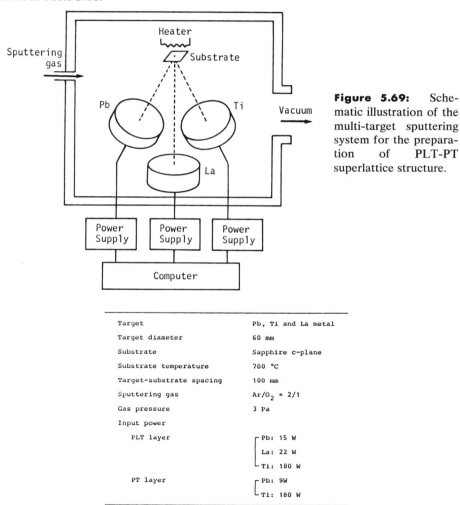

Figure 5.69:   Schematic illustration of the multi-target sputtering system for the preparation of PLT-PT superlattice structure.

Target	Pb, Ti and La metal
Target diameter	60 mm
Substrate	Sapphire c-plane
Substrate temperature	700 °C
Target-substrate spacing	100 mm
Sputtering gas	$Ar/O_2 = 2/1$
Gas pressure	3 Pa
Input power	
PLT layer	Pb: 15 W La: 22 W Ti: 180 W
PT layer	Pb: 9W Ti: 180 W

Table 5.18:   Typical sputtering conditions for the deposition of PLZT thin films by a multi-target deposition system.

Note that the input power of the Pb target for depositing the PLT layer is larger than for the PT layer despite the fact that the relative composition of Pb in PLT is smaller than in PT. This phenomena is due to the re-evaporation of Pb occurring during the growth of PLT due to the weak affinity between Pb and La atoms. Thus excessive incident Pb content is required. The deposition times for PLT and PT layers are kept equal and the ratio of thickness (PLT/PT) is about 3/2 at these conditions. The superlattice films with a total thickness of about 3000Å are prepared by varying the period of PLT-PT deposition from 120 to 300Å/sec.

Reflection high-energy electron diffraction (RHEED) analysis suggests that the sputtered films are epitaxially grown with the relationship, (111) perovskite ‖ (0001) sapphire and [10$\bar{1}$] perovskite ‖ [10$\bar{1}$0] sapphire. Figure 5.70 shows the X-ray diffraction pattern of a film grown at a deposition period of 300 sec. Diffraction from (111) plane of perovskite structure with a pair of satellite peaks are observed. This suggests the presence of a superlattice structure in the sputtered film.

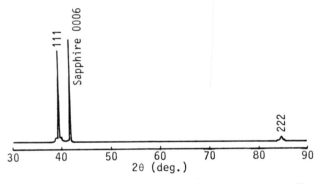

Figure 5.70: X-ray diffraction pattern for the PLT-PT superlattice film on sapphire grown at the deposition period of 300 sec.

Figure 5.71 shows the X-ray diffraction patterns of (111) peak as a function of deposition period. When the deposition period is as long as 3000 sec, the (111) peak is divided into (111) peaks of PLT layer and PT layer as shown in Fig. 5.71(a). However, for the film grown at the deposition period of 600 sec, a new strong peak appears between PLT (111) and PT (111) peak, and some satellite peaks around the center peak are observed as seen in Fig. 5.71(b). These satellite peaks are the first (Θ_1^+, Θ_1^-) and second (Θ_2^+, Θ_2^-) diffracted peaks caused by the superlattice structure with a modulation wavelength of 330Å. The modulation wavelength Λ is calculated by the equation,

$$\Lambda = \frac{\lambda}{\sin \Theta_i^+ - \sin \Theta_i^-}, \tag{5.8}$$

where λ is the wavelength of the X-ray and Θ_i^+, Θ_i^- are the Bragg angles of ith satellite peaks. As the deposition period becomes shorter, the wavelength becomes shorter and the intensity of the satellite peaks decreases. For films grown at a deposition period of 120 sec, no satellite peaks are observed. This result indicates that inter-diffusion of each layer occurs and the modulation structure disappears. The relation between the deposition period and the modulation wavelength is plotted in Fig. 5.72. Clear linearity is observed in the figure. The modulation wavelength can be strictly determined by the deposition period. These results suggest that the multi-target sputtering system is useful for the deposition of the ferroelectric compound thin films with a controlled chemical composition and crystal structure. Figure 5.73 shows a typical SEM image and RHEED pattern of an epitaxial $(Pb_{0.77}La_{0.23})Ti_{0.94}O_3$ thin film prepared by the multi-target sputtering system.

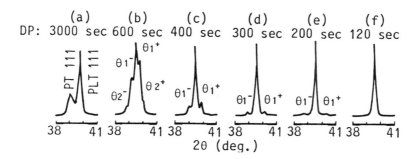

Figure 5.71: X-ray diffraction patterns of (111) peak for the films grown at the various deposition period (DP).

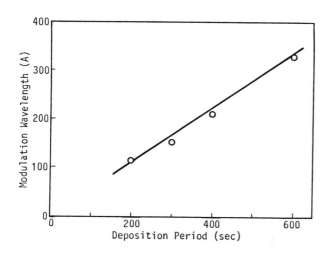

Figure 5.72: The relation between the deposition period and the modulation wavelength.

The transmission spectra of the PLZT (28/0/100) films are shown in Fig. 5.74. The films are transparent from visible to the near-infrared region. The refractive indexes of PLZT films varied with the Pb content in the film and was in the range of 2.4 - 2.7 at $0.633\,\mu m$.

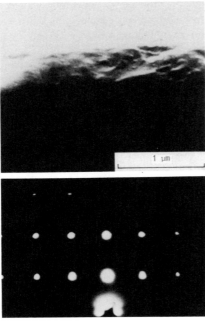

Figure 5.73: Typical SEM and RHEED image of $(Pb,La)TiO_3$ thin film.

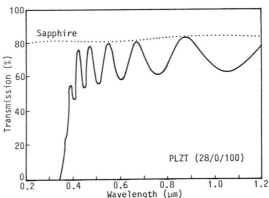

Figure 5.74: Optical transmission spectrum of epitaxial PLZT (28/0/100) films about 0.4 μm thick.

Electrical Properties: Dielectric properties of the sputtered PLZT have been measured with the sandwich structure shown in Fig. 5.75. The PLZT thin films are deposited on a sapphire substrate overcoated by a TiN thin film electrode. The sputtered films exhibit polycrystalline form and show the dielectric anomaly similar to ceramics. However, the broad transition is observed in the temperature-permittivity characteristics as shown in Fig. 5.76.

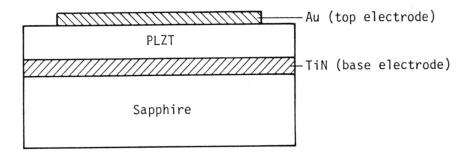

Figure 5.75: Structure of electrodes for the measurements of dielectric properties of PLZT thin films.

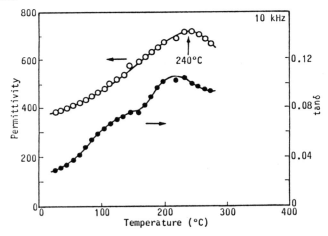

Figure 5.76: Temperature dependence of dielectric properties for PLZT (9/65/35) film.

Dielectric properties of epitaxial PLZT thin films are evaluated by the deposition of comb Al electrodes on the surface of the PLZT thin films as shown in Fig. 5.77. In the structure, the thin-film dielectric constant of is calculated from the measured capacitance C of the Al comb electrodes on top of the film using the following approximation (131):

$$C = Knl[(\varepsilon_s + 1) + (\varepsilon_f - \varepsilon_s)[1 - \exp(-4.6h/L)]], \tag{5.9}$$

where ε_s is the dielectric constant of the substrate ($\varepsilon_s = 10$), h is the thin-film thickness (h = 0.33μm), L is the center-to-center spacing between adjacent electrodes (L = 6μm), n is the number of electrode strips (n = 160), l is the length of the fingers of electrodes (l = 720x10^{-6}m), and K is the constant given by the structure of electrodes (K = 4.53x10^{-12}).

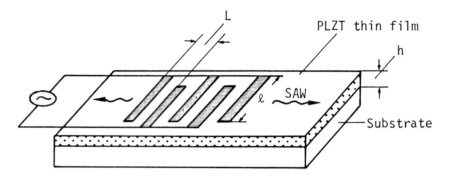

Figure 5.77: Comb electrodes for the measurement of dielectric properties of epitaxial PLZT thin films.

Figure 5.78 shows the temperature dependence of the relative dielectric constant measured for films with various compositions. The peaks of the dielectric constants correspond to the Curie temperatures (T_c), since the D-E hysteresis measured using a Sawyer-Tower circuit disappeared at temperatures above the peaks (132). Typical D-E hysteresis curves of PLZT thin films measured at various temperature are shown in Fig. 5.79. Note that the dielectric constant maximum for PLZT (9/65/35) film shows a broad temperature dependence which is similar to that of PLZT (9/65/35) ceramic, but the T_c of the film is approximately 100°C higher than that of the ceramic. Table 5.19 indicates the dielectric properties of PLZT thin films compared with those of ceramics of various chemical composition.

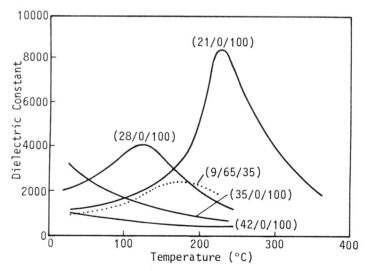

Figure 5.78: Temperature dependence of the relative dielectric constant of PLZT films about 0.4μm thick. The measuring frequency is 100 kHz.

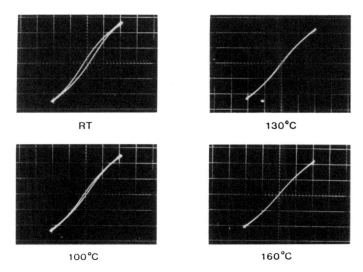

| RT | 130°C |
| 100°C | 160°C |

Figure 5.79: Temperature dependence of D-E hysteresis loop of PLZT (28/0/100) thin film at 60 Hz. Horizontal scale: 5 V/div. Vertical scale: 3×10^{-9} C/div.

| Target Composition | Thin Films | | Ceramics | |
	ϵ^*	T_c (°C)	ϵ^*	T_c (°C)
PbTiO$_3$	370	490	230	490
PLZT (0/65/ 35)	450	275		365
PLZT (7/65/ 35)	480	260	1570	150
PLZT (9/65/ 35)	710	240	4650	85
PLZT (11/65/ 35)	630	220	4100	70
PLZT (14/65/ 35)	380	220	1450	50
PLZT (14/ 0/100)	600	290	1200	220
PLZT (21/ 0/100)	1300	225	2000	100
PLZT (28/ 0/100)	1800	120	2000	-100
PLZT (42/ 0/100)	1100			

* measured at 10 kHz, RT.

Table 5.19: Dielectric properties of PLZT thin films prepared by sputter deposition.

It is understood that the T_c of PLZT (x/0/100) ceramics will increase with the decrease of La content (133). Similar phenomena have been observed in sputtered PLZT thin films. The T_c of PLZT (x/0/100) films and ceramics are compared in Fig. 5.80. The T_c of thin films are higher than those of ceramics. Under the assumption that the relationship between T_c and the La content of the thin films agrees well with that of ceramics, the disagreement of T_c between films and ceramics may be due to the compositional difference of the thin films. These assumptions are confirmed by chemical analyses of the sputtered PLZT thin films.

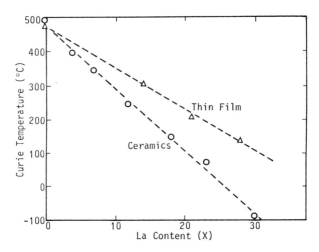

Figure 5.80: Curie temperature vs La content of ceramics and thin films deposited by PLZT (x/0/100) targets.

Their piezoelectric properties are evaluated by excitation of the SAW. An as-grown state of epitaxial PLZT thin film has three equivalent anisotropic axes along the edges of a pseudo-cubic lattice, which makes an angle of about 35 degrees to the film plane. Poling treatment is done as follows: First, Al electrodes are fabricated with 1 mm gaps on the film surface. Then the temperature is elevated to 200°C which is higher than Curie temperature. Next, the sample is gradually cooled from 200°C by applying a voltage of 2 kV as shown in Fig. 5.81(a). The direction of the applied electric fields is parallel to ($\overline{1}12$) of PLZT. The polarization of this region in the thin film will be uniformly arranged to (001). In order to excite and detect SAW, interdigital transducers (IDT) of Al are made by a lift-off method on the polarized regions as shown in Fig. 5.81(b). The period of the IDT finger is 12μm and the pair number of the fingers is 80. SAWs will propagate along the <$\overline{1}12$> direction of PLZT.

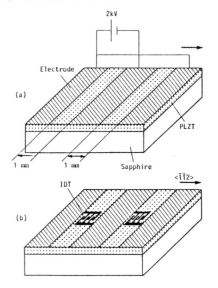

Figure 5.81: Fabrication procedure of PLZT SAW devices.

Figure 5.82 shows the Smith chart plot of the impedance characteristic of IDT. The normalized thickness of the film is Kd = 0.42, where K is the wave number of SAW and d is the film thickness. Two modes exist whose center frequencies are 405 MHz and 455 MHz. These modes are the fundamental (0th) and the higher-order modes of SAW, respectively. The electromechanical coupling constant k^2 is evaluated by Smith's relationship

$$k^2 = \frac{\pi^2 f_o C_T R_q}{2N},\qquad(5.10)$$

where N is the pair number of the fingers of IDT, C_T is the capacitance of IDT, f_o is the center frequency and R_q is the measured radiation resistance (134). The coupling constant k^2 is calculated to be about 0. 85% for the 0th mode of SAW, which is a relatively large value.

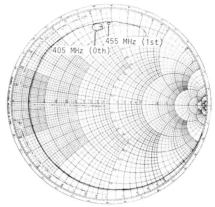

Figure 5.82: Smith chart impedance pattern for the IDT fabricated on the PLZT thin film. Frequency range is 350-450 MHz.

The phase velocity V_p and the coupling constant k^2 of SAW for various PLZT thicknesses are shown in Fig. 5.83. The open circle indicates the 0th mode SAW and the solid circle indicates the 1st mode. Since the physical parameters of the present PLZT are unknown, we have calculated the SAW properties using the parameters of BaTiO$_3$ and PbTiO$_3$ for comparison (135,136). These are similar perovskite-type ferroelectric materials. The epitaxial relationships and the propagating direction of SAW were similar to the experiment. The coupling constant k^2 is expressed as follows

$$k^2 = 2F \frac{\Delta V}{V_p},\qquad(5.11)$$

where V_p is the SAW velocity, ΔV is the perturbation of velocity, and F is the filling factor. The calculation is carried out substituting 1 for F as usual. The results are shown in the figure by solid lines for BaTiO$_3$ and broken lines for PbTiO$_3$. Although the composition of the present PLZT is similar to PbTiO₃, the SAW properties show characteristics close to BaTiO₃. The piezoelectric effect of the PLZT thin film seems as strong as that of BaTiO₃.

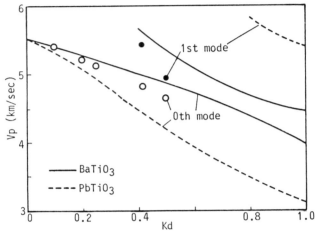

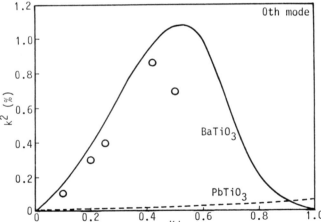

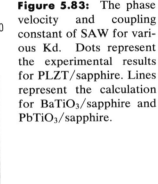

Figure 5.83: The phase velocity and coupling constant of SAW for various Kd. Dots represent the experimental results for PLZT/sapphire. Lines represent the calculation for BaTiO₃/sapphire and PbTiO₃/sapphire.

The PLZT films also show high electro-optic properties. Figure 5.84 shows a typical birefringence shift as a function of the electric field measured by an elipsometic method. The characteristic shows a nearly quadratic effect with small hysteresis and the birefringence shift reaches $\Delta n \simeq -0.0015$ at an applied electric field of 2 kV/m. The electro-optic coefficient R is $0.8 \times 10^{-16} (m/V)^2$ under the assumption that the $\Delta(\Delta n) - E$ characteristic exhibits a quadratic curve governed by the Kerr effects. The electro-optic coefficient is almost the same as those of PLZT bulk ceramics of similar composition (126). Table 5.20 shows summaries of the electrical properties of the PLZT thin films.

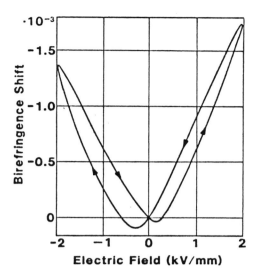

Figure 5.84: Typical birefringence shift as a function of electric field for the PLZT(28/0/100) thin film about 0.83μm thick.

Optical switches: Light beam switching in a four-port channel waveguide has has been achieved using an electro-optic modulation of epitaxially grown PLZT thin film on sapphire (137).

Figure 5.85 shows the typical configuration of channel waveguide switches. The switches are composed of a total internal reflection (TIR) structure (138). Thickness of the PLZT thin film is 3500Å. The PLZT thin film is prepared by rf-planar magnetron sputtering from a target of sintered PLZT (28/0/100) powder. The four-port channel waveguides are formed by ion beam etching. The intersecting angle is 2.0°. The width of each channel waveguide is 20μm. A pair of parallel metal electrodes separated 4μm apart from each other are deposited at the center of the intersection region. In the absence of a switching voltage, incident guided-light beam from port 1 encounters no refractive index fluctuation between the parallel electrodes and propagates straight to the port 3. When the switching voltage performs due to a refractive index fluctuation, the guided-light beam is reflected to port 4.

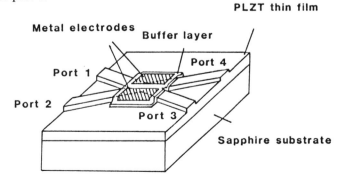

Figure 5.85: Configuration of optical TIR switches comprising PLZT thin film/sapphire layered structure.

Materials	Deposition method* (source)	Substrate	Deposition temp. (°C)	Structure**	Film properties	Ref.
PZT	EB	SiO_2	350 (postanneal at 700°C)	PC	$\epsilon^* \sim 100$ (RT) $P_s \sim 4.2$ $\mu C/cm^2$ $T_c \sim 340$ °C	a
	RF-SP (PZT 52/48)	SiO_2	> 500	PC	$\epsilon^* \sim 751$ (RT) $P_s \sim 21.6$ $\mu C/cm^2$ $T_c \sim 325$ °C, $n_0 = 2.36$	b
PLT	RF-SP (PLT 18/100)	MgO	600 $\sim$ 700	SC	$\epsilon^* \sim 700$ (RT) $n_0 = 2.3 \sim 2.5$ (6328Å)	c
	RF-MSP (PLT 28/100)	sapphire	580	SC	$\epsilon^* \sim 2000$ (RT) $T_c \sim 150$ °C $n_0 = 2.4 \sim 2.7$ (6328Å) electro-optic coefficient $R \sim 0.6 \times 10^{-16} m^2/v^2$ (6328Å) SAW coupling $k_{SAW}^2 = 0.85\%$ (Kd=0.4)	d
PLZT	RF-SP (7/65/35)	SiO_2	500 (postannealing 650 $\sim$ 700°C)	PC	$\epsilon^* = 1000 \sim 1300$ $T_c \sim 170$ °C	e
	RF-SP (9/65/35)	sapphire $SrTiO_3$	700	SC	$n_0 \simeq 2.49$ (6328Å)	f
	RF-MSP (9/65/35)	sapphire	580	SC	$\epsilon^* \sim 710$ $T_c \sim 240$ $R \sim 1 \times 10^{-16} m^2/v^2$ (6328Å)	d

*   EB, electron beam deposition; SP, diode sputtering; MSP, magnetron sputtering

**   PC, polycrystalline; SC, single crystal

Table 5.20: Properties of PZT, PLZT thin films.

REFERENCES for Table 5.20:

a. Oikawa, M., and Toda, K., Appl. Phys. Lett., 29: 491 (1976).
b. Okada, A., J. Appl. Phys., 48: 2905 (1977).
c. Usuki, T., Nakagawa, T., Okumuyama, M., Karaya, T., and Hamakawa, Y., 1978 Fall Meeting of Applied Phys, Japan, paper 3p-F-10 (1978).
d. Adachi, H., Mitsusy, T., Yamazaki, O., and Wasa, K., J. Appl. Phys., 60: 736 (1986).
e. Nakagawa, T., Yamaguchi, J., Usuki, T., Matsui, Y., Okuyama, M., and Hamakawa, Y., Jpn. J. Appl. Phys., 18: 897 (1979).
f. Ishida, M., Tsuji, S., Kimura, K., Matsunami, H., and Tanaka, T., J. Crystal Growth, 45: 393 (1978).

Photographs of the transmitted and reflected light beam are shown in Fig. 5.86. The switching speed is higher than GHz with a switching voltage of less than 5V (139). These types of switches can be used for high-speed multiplexers and demultiplexers for an optical LAN system (140).

Since a variety of thin-film optical devices including lasers, light detectors, acoustic deflectors, and other micro- optical elements can be deposited on the same sapphire substrates, the PLZT/sapphire layered structure has the potential for making novel integrated optic circuits (141).

Reflected   Transmitted

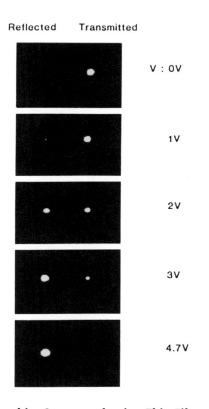

V : 0V

1V

2V

3V

4.7V

Figure 5.86: Transmitted (port 3) and reflected light beam intensity (port 4) at the PLZT thin film TIR switch.

5.1.4 Perovskite Superconducting Thin Films

Extensive work has been done on high-temperature superconducting conducting ceramics since Bednorz and Müller discovered the La-Ba-Cu-O compound system with the transition temperature of $T_c \approx 30K$ (142). The K_2NiF_4 structure is found to be the major phase. Numerous compositions have been studied in an effort to raise the T_c. The replacement of Ba ions by smaller Sr ions in the La-Ba-Cu-O compound system has elevated the T_c (143). The oxygen-deficient perovskite, Y-Ba-Cu-O system with $T_c = 90K$ has been developed by Chu (144). These compounds are composed of rare earth elements. Extensive research on superconductors has led to several new high T_c oxide materials, including rare earth-free high T_c oxides of Bi-Sr-Ca-Cu-O and Tl-Ba-Ca-Cu-O systems with T_c exceeding 100 K developed by Maeda and Hermann respectively

(145,146).   These high temperature oxide superconductors which are composed of a copper oxide layer are shown in Table 5.21.

Historically oxide superconductors have been known since 1964. Table 5.22 shows these oxide superconductors. These oxides are composed of perovskite $SrTiO_{3-\delta}$ with a low transition temperature of 0.55 K. The history of high T_c superconductors of the perovskite goes back to $BaPb_{1-x}Bi_xO_3$ with a T_c of 13 K proposed by A. W. Sleight in 1974 (147).

A_2BO_4 $(La_{1-x}M_x)_2CuO_4$	M: Ca, Sr, Ba	Tc: 20-40 K
$A_3B_3O_{7-\delta}$ $Ba_2LnCu_3O_{7-\delta}$	Ln: Y, Nd, Sm, Eu, La, Lu, Gd, Dy, Ho, Er, Tm, Yb	Tc: 90 K
$T_2A_2BO_x$, $T_2A_3B_2O_y$, $T_2A_3B_3O_z$	T: Bi, A: Sr,Ca, B: Cu T: Tl, A: Ba,Ca, B: Cu	Tc: 80-120 K
Miscellaneous	$(Nd_{0.8}Sr_{0.2}Ce_{0.2})_2CuO_4$ $(Nd_{1-x}Ce_x)_2CuO_4$ $(x{\sim}0.07)$ $(Ba_{1-x}M_x)BiO_3$ $(x{\sim}0.4)$ M=K,Rb	Tc: 27 K 25 K 30 K

Table 5.21: High T_c superconducting oxides.

Materials	Tc	Structure	Date
$SrTiO_{3-\delta}$	0.55 K	perovskite	1964
TiO	2.3	NaCl	1964
NbO	1.25	NaCl	1964
M_xWO_3	6.7	tungsten bronze	1964
$Ag_7O_3{}^+X^-$ $(X^-=NO_3{}^-, F^-, BF_4{}^-)$	1.04	clathrate	1966
$Li_{1+x}Ti_{2-x}O_4$	13.7	spinel	1973
$BaPb_{1-x}Bi_xO_3$	13	perovskite	1975

Table 5.22: Traditional oxide superconductors.

The crystal structures of high T_c perovskite-related copper oxide superconductors are shown in Fig. 5.87. In the perovskite compound ABO_3, the A-site and the B-site elements

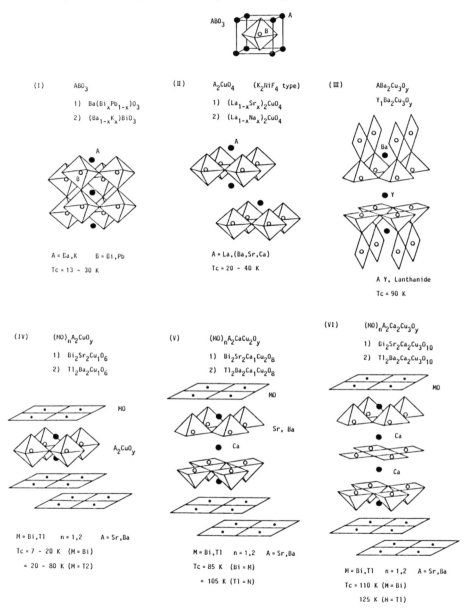

Figure 5.87: Crystal structures of oxide superconductors.

could be substituted by other kinds of cations. High oxidizations are achieved by substituting the B-side with 3d-transition metals accompanied by the generation of a mixed-valence state. These features are observed in perovskite-related high T_c superconductors.

The basic properties of copper oxide superconductors have been reviewed by several workers (148) and are shown in Table 5.23. Note that this type of oxide often shows a large oxygen nonstoichiometry. Figure 5.88 shows the $\delta - \log P(O_2)$ curves for various ambient temperatures in the $A_3B_3O_{7-\delta}$ system (149). It is seen that the superconducting orthorhombic phase is stable at temperatures below 600 - 650°C.

Material	T_c (K)	ξ (nm)	Λ (nm)	κ (Λ/ξ)	
LaSr	37		250		Aeppli(1987)
	36	3	330	110	Finnemore(1987)
		1.8			Kobayashi(1987)
	34		200		Kossler(1987)
	38	1.3	210	160	Orlando(1987)
				~75	Renker(1987)
				~40	Takagi(1987)
	37		230		Wappling(1987)
LaSr(0.096)	38	2.0			Nakao(1987)
LaSr(0.1)		2.6			Kobayashi(1987)
	35	1.3	210	160	Orlando(1987)
	39	2.0	100	50	Uchida(1987)
LaSr(0.3)	35	3.2			Murata(1987)
YBa				180	Bezinge(1987)
			22.5		Felici(1987)
		1.5·	120	80	Gottwick(1987)
				70	Grant(1987)
	95	2.7($\parallel$)			Hikita(1987)
	95	0.6($\perp$)			Hikita(1987)
	84		130		Kossler(1987)
	89	1.7			Orlando(1987)
	89	3.4($\parallel$)	26($\perp$)	7.6($\parallel$)	Worthington(1987)
	89	0.7($\perp$)	125($\parallel$)	37($\perp$)	Worthington(1987)
			400		Zuo(1987)
$YBa_2Cu_3O_{6.9}$	92.5	2.2	140	65	Cava(1987)
$Y_{0.4}Ba_{0.6}CuO_{3-\delta}$	89	1.4			Murata(1987)
BiSrCaCuO		4.2($\parallel$)			Hidaka(1988)
		0.1($\perp$)			Hidaka(1988)

Coherence Lengths (ξ), Penetration Depths (Λ), and Their Ratios   (Ginzburg–Landau Parameter) $\kappa = (\Lambda/\xi)$

The notation used is: $(La_{1-x}M_x)_2CuO_{4-\delta}$ = LaM(x); LaM (0.075) = LaM$_\bullet$; $YBa_2Cu_3O_{7-\delta}$ = YBa$_\bullet$.
Several values of the coherence length $\xi_\parallel = \xi_{ab}$ in the Cu-O planes and $\xi_\perp = \xi_c$ perpendicular to these planes are given.

Table 5.23: Material parameters for high T_c superconductors (Pool, 1988, (148)).

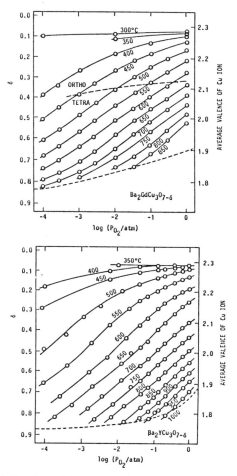

Figure 5.88: Oxygen deficiency and average valence of Cu ion as a function of log P(O₂) (Fueki (1988) (150)).

5.1.4.1 Studies of Thin Film Processes: Thin films of perovskite-related high T_c superconductors have been successfully synthesized. This is due to previous studies on dielectric thin films of perovskite-type oxides (39) and/or superconducting thin films of the perovskite-type, $BaPb_{1-x}Bi_xO_3$ (59), and the alloy superconductors of the $A_{15}Nb_3Ge$ and B_1 NbN type (51,53).

In the early periods of research on these thin films, sputtering and/or electron beam deposition was used for the preparation of the thin films. The polycrystalline and/or single crystal thin films of the La-Sr-Cu-O were prepared by several workers (60-64). Adachi found that sputtered single crystal thin films of La-Sr-Cu-O exhibited a transition temperature $T_c = 34K$ which corresponded to the best results for the ceramics (63). Suzuki found that this kind of the superconductor shows a small carrier density of around $n_H = 6.8 \times 10^{21} cm^{-3}$ similar to the $BaPb_{1-x}Bi_xO_3$ oxides (61). Naito evaluated the energy gap of La-Sr-Cu-O thin films by infrared reflectance measurements; the values were found to be 20-30 mV (64).

After the discovery of Y-Ba-Cu-O high T_c superconductors by Chu (145), most of thin film studies were shifted to the deposition of Y-Ba-Cu-O thin films. The sputtered films exhibited high transition temperature of around 90K which was close to the value of ceramics. Their critical currents were found to exceed $10^6 A/cm^2$ at liquid nitrogen temperature of 77 K under zero magnetic field, although critical currents measured in bulk ceramics were less than $10^3 A/cm^2$ (67).

Several basic properties in the new high T_c Y-Ba-Cu-O thin films were evaluated, including the tunneling gap and anisotropy in critical currents and critical field (67,150). In these experiments the critical field was found to exceed 4 T/K (150). These extensive studies have suggested that the new high T_c superconductors have possible applications for electronic devices and/or power systems.

The possibility of lowering the synthesis temperature has been discussed. Sputter deposition has allowed the reduction of the synthesis temperature from $900°C$ to $600°C$ by irradiation with an oxygen plasma during deposition (69). Lowering the synthesis temperature has been found to reduce the mutual diffusion between thin films and the substrates, and stabilizes the interface (151). Several workers have studied mutual diffusion between thin films and substrates (152,153). Al atoms in the sapphire substrates are found to easily diffuse into superconducting thin films during post annealing. Buffer layers are used for the reduction of mutual diffusions. ZrO_2 (154), CaF_2 (151), and Pt (155) layers are considered as buffer layers on Si and sapphire substrates.

A multi-layer deposition was also examined to keep the correct film stoichiometry (156). A new deposition process, pulsed laser deposition, has also been evaluated for controlled deposition of high T_c superconductors (157). Chemical vapor deposition is also considered as an available method for deposition of new high T_c superconductors (158).

Thin film deposition of the high T_c superconductors of Bi-Sr-Ca-Cu-O and/or Tl-Ba-Ca-Cu-O systems have been tried similar to La-Sr-Cu-O and/or Y-Ba-Cu-O thin films. These deposition methods are listed in Table. 5.2.

In these deposition methods the most important problem is to keep the correct composition. The layer-by-layer deposition in an atomic scale proposed by Adachi is one of the most promising methods for the controlled deposition of high T_c superconductors (159).

Aside from deposition, extensive studies on passivation and/or microfabrication have been done by several workers (160,161). Hirao has found that ECR plasma CVD can be used for making a stable passivation layer onto high T_c superconducting thin films due to low working pressure and low deposition temperatures (162).

5.1.4.2 Basic Thin Film Processes: Basic processes for deposition of perovskite thin films are shown in Figure 5.89. Thin films of amorphous phase are deposited at the substrate temperature T_s below the crystallizing temperature T_{cr}, which is $500 - 700°C$ for the perovskite type oxides. In some cases a different crystal structure appears at substrate temperatures below the T_{cr} for perovskite structure. For thin films of $PbTiO_3$ the pyrochlore phase appears at substrate temperatures around $400°C$ (163).

```
1.  AMORPHOUS PHASE              Ts < Tcr

2.  POLYCRYSTALLINE             Ts > Tcr
                                Ts < Tcr, postannealing

3.  SINGLE CRYSTALS             Ts > Te
    (single crystal sub.)       Ts < Tcr, postannealing
                                (solid-phase epitaxy)
```

Figure 5.89: Basic process of the deposition of the perovskite thin films: T_s, substrate temperature during deposition; T_{cr}, crystallizing temperature during deposition; T_e, epitaxial temperature.

Thin films of polycrystalline phase are deposited at $T_s > T_{cr}$ This phase is also achieved by deposition of the amorphous phase followed by post annealing at temperatures above T_{cr}.

Thin single crystal films are epitaxially deposited on a single crystal substrate at substrate temperatures above the epitaxial temperature T_e ($T_e > T_{cr}$). The amorphous thin films deposited on the single crystal substrate will be converted into single crystalline thin films after post annealing at temperatures above T_e owing to solid phase epitaxy.

Several kinds of deposition processes are proposed for perovskite type oxides, including electron beam deposition (164), laser beam deposition (165), cathodic sputtering (166), and chemical vapor deposition (165). Oxidization is considered necessary for deposition of high T_c superconductors. For this purpose oxygen, ozone and/or oxygen ions are supplied onto the growing surface of thin films during the deposition by the electron beam and/or molecular beam deposition system as shown in Fig. 5.90.

In cathodic sputtering, thin films of perovskites are deposited directly from the compound ceramic target of perovskites in an rf-system. Rf-magnetron sputtering is commonly used for deposition from the compound ceramic target. A sintered ceramic plate or sintered ceramic powder is used for the sputtering target. In dc magnetron sputtering, the metal targets of A-site and B-site elements are sputtered in an oxidizing atmosphere.

Wehner and coworkers have described that in a planar diode system the film composition from a multi component target is a function of substrate location, and is usually different from that of the target. The main reason for this is that different atomic species are sputtered with different angular distributions. An additional problem arises with negative oxygen ions which cause presputtering from substrates located opposite the target. These complications disappear when sputtering is performed using spherical targets (168) although this is generally an impractical means of film formation.

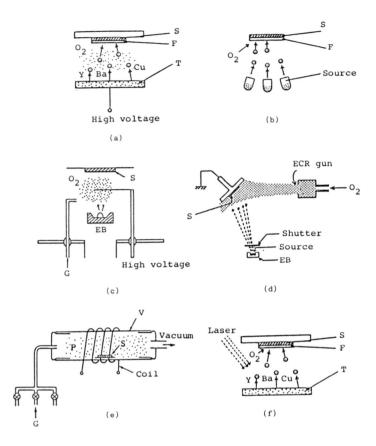

Figure 5.90: Typical deposition systems for the high T_c superconductors; (a) sputtering, (b) reactive evaporation, (c) activated reactive evaporation, (d) ion assisted evaporation, (e) plasma CVD, (f) reactive laser abrasion. S: substrate, F: thin film, T: target, ES: evaporation source, V: vac. chamber, EB: electron beam, G: reactive gas source.

In chemical vapor deposition metal-organic compounds such as $Y(C_{11}H_{19}O_2)_3$, $Ba(C_{11}H_{19}O_2)_2$, and $Cu(C_{11}H_{19}O_2)_2$ are tentatively used as the source for deposition of YBC thin films. Halides such as $BiCl_3$, CuI, CaI_2, and SrI_2 are used as the source for deposition of BSCC thin films (169).

For the deposition of single crystal films, the selection of the substrate crystal will affect the crystal properties of the resultant films. The crystallographic properties of the crystal substrates used for epitaxial growth of perovskite type oxides are shown in Table 5.24 (170).

	Crystal System	Structure	Lattice Constants (A)	Thermal Expansion Coefficient (K^{-1})
$La_{1.8}Sr_{0.2}CuO_4$	tetragonal #	K_2NiF_4	a=3.78 c=13.23	$10-15 \times 10^{-6}$
$YBa_2Cu_3O_x$	orthorhombic	oxygen deficient perovskite	a=3.82 b=3.89 c=11.68	$10-15 \times 10^{-6}$
Bi-Sr-Ca-Cu-O	pseudo tetra.	Bi-layered structure	a=5.4 c=30,36	12×10^{-6}
Tl-Ba-Ca-Cu-O	pseudo tetra.	Bi-layered structure	a=5.4 c=30,36	
sapphire (α-Al_2O_3)	trigonal	corundum	hex. axes a'=4.763 c'=13.003	$\parallel$c 8×10^{-6} $\perp$c 7.5×10^{-6}
MgO	cubic	NaCl	a=4.203	13.8×10^{-6}
$MgAl_2O_4$	cubic	spinel	a=8.059	7.6×10^{-6}
YSZ	cubic	fluorite	a=5.16	10×10^{-6}
$SrTiO_3$	cubic	perovskite	a=3.905	10.8×10^{-6}
$LaGaO_3$	orthorhombic	perovskite	a=5.482 b=5.526 c=7.780	9×10^{-6}
$LaAlO_3$	pseudo cubic	perovskite	a=3.792	10×10^{-6}
$Nd:YAlO_3$	orthorhombic	perovskite	a=5.18 b=5.33 c=7.37	2.2×10^{-6}
$NdGaO_3$	orthorhombic	perovskite	a=5.426 b=5.502 c=7.706	7.8×10^{-6}

\# Superconductivity at orthorhombic phase.

Table 5.24: Substrates for the deposition of the high T_c superconducting thin films.

Figure 5.91 shows a typical epitaxial relationship between high T_c superconductors and cubic substrates. The c-axis of the epitaxial films will be perpendicular to the (100) plain of the substrate crystals. The isotropic superconducting currents will flow in the (001) plain of the deposited films. On the (110) plain of the cubic crystal substrates the c-axis of the epitaxial films will lie in the films. Large anisotropy will be expected for the current flow in the (110) plain of the epitaxial films.

It should be noted that most of the crystal substrates exhibit cubic structure. The crystal twin will often be formed in epitaxial films of orthorhombic superconductors, and orthorhombic substrates are important for the reduction of them. Besides crystallographic properties, the possibility of mutual diffusion at the film and substrate interface should be considered in the selection of the substrates (171,172).

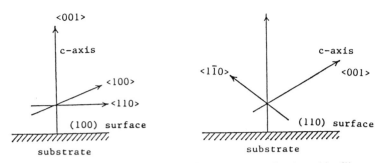

Figure 5.91: Epitaxial relations of the high T_c superconducting thin films on crystal substrate.

A number of experiments have been done for deposition of high T_c thin films. For the Y-Ba-Cu-O system three deposition processes are classified as indicated in Table 5.25.

		Chemical Composition	Crystallization	Oxygen Vacancy and/or structural Control
Ceramics		mixing	sintering (850 - 950°C)	annealing * (850 - 950°C)
Thin films	1	deposition (Ts > Tcr)	annealing (850 - 950°C)	annealing * (850 - 950°C)
	2	deposition (Ts > Tcr)		annealing * (400 - 950°C)
	3	deposition ** (Ts > Tcr)		

Ts: substrate temperature during deposition   * slow cooling
Tcr: crystallizing temperature (500 - 700°C)   ** quenching

Table 5.25: Fabrication processes for high T_c superconducting ceramics and thin films.

The process for making YBC ceramics is composed of three stages; (i) mixing, (ii) annealing for crystallization and sintering, and (iii) annealing for the control of oxygen vacancies and/or crystal structure. In the annealing process, the oxidation of copper will be promoted and the density of Cu^{+1} and/or Cu^{+2} decrease and Cu^{+3} density increases.

For YBC thin films, three processes are considered; process (1), deposition at low substrate temperature followed by postannealing. This process (1) is commonly used for deposition of high T_c superconducting thin films, since stoichiometric composition of the thin films is relatively easily achieved. Single crystal thin films are expected to be obtained under the condition of solid phase epitaxy during the postannealing process. At present, however, the resultant films show the polycrystalline phase. Process (2) is deposition at high substrate temperature followed also by a post-annealing process. Process (3) is the deposition at high substrate temperature without a subsequent post-annealing process. Single crystal films can be obtained by vapor phase epitaxy achieved in process (2) and (3). These considerations may also be adopted for deposition of the LSC system and the BSCC and/or the TBCC system. However, in the LSC system, the oxidation of copper will be promoted by the substitution of La^{+2} site by Sr^{+3} during the postannealing process. In the BSCC and/or TBCC system, the rearrangements of Sr,Ba, and/or Ca will act the oxidation as described later.

5.1.4.3 Synthesis Temperature:
As indicated in Table 5.25 the synthesis temperature of high T_c superconducting ceramics is around $850 - 950°$. Lowering it is very important not only for scientific interests but also for fabrication of thin film superconducting devices. Table 5.25 shows that the maximum temperature in the thin film process may be governed by the postannealing process. In thin film process (2) for rare earth $YBa_2Cu_3O_x$ superconductors, the synthesis temperature is governed by the postannealing process for control of the oxygen vacancies, if the as-deposited thin films are crystallized.

The structural analysis for $YBa_2Cu_3O_x$ ceramics suggest that the structural transition from the non-superconducting tetragonal phase ($x \simeq < 6.3$) to the superconducting orthorhombic phase ($7 > x > 6.3$) occurs around $700°C$, and the latter phase is predominant at the annealing temperatures below $600°C$ as shown in Figure 5.88 (173). It is known that when the annealing temperature is near the tetra/ortho transition temperature, superconductors show the ortho-II phase with $T_c = 50 - 60K(6.7 \geq x > 6.3)$. The ortho-I phase with $T_c = 90K(7 > x > 6.7)$ is obtained at a lower annealing temperature below $600°C$ (174).

Similar results are obtained in Gd-Ba-Cu-O (GBC) films (175). Figure 5.92 shows typical experimental results on the variations of the c-axis lattice parameter with the postannealing temperature for the c-axis oriented GBC films of the tetragonal phase, $c = 11.83 - 11.84Å$. It shows that the c-axis lattice constant is reduced by post annealing in O_2. The films annealed at $550 - 650°C$ with $c = 11.73 - 11.75Å$ may correspond to the ortho-II phase. The high T_c ortho-I phase with $c = 11.71 - 11.73$ Å and $T_c = 90$ K is obtained at an annealing temperature of $350 - 550°C$. These structural analyses suggest that the synthesis temperature is not governed by the postannealing temperature for YBC films but by the crystallizing temperature.

For rare-earth $La_{2-x}Sr_xCuO_4$ thin films the as-deposited films show the insufficient oxidation of copper. Oxidation takes place during the post annealing process. The maximum synthesis temperature for LSC thin films may also be governed by the crystallization temperature since oxidation will take place below it. The crystallization temperature for these rare earth high T_c oxides is around $500 - 600°C$. For rare earth free high T_c superconductors, the minimum synthesis temperature may also correspond to the crystallization temperature of perovskites, $500 - 700°C$.

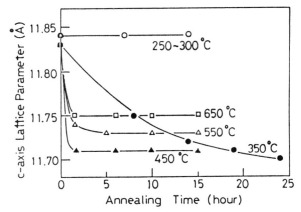

Figure 5.92: Variation of the c-axis lattice parameter of the Gd-Ba-Cu-O thin films with postannealing in O_2 atmosphere at various temperatures.

Note that in YBC thin films the partial substitution of O- sites by S shows a tendency to decrease the crystallizing temperature accompanied by a sharpened transition (176,177).

5.1.4.4 Low Temperature Processes/In Situ Deposition: It is possible that lowering the synthesis temperature the YBC system for the YBC system can be achieved by deposition at the crystallizing temperature of $500 - 700°C$ followed by post annealing in O_2 at temperature of $350 - 550°C$. The maximum temperature for rare earth high T_c film processes is then governed by a crystallizing temperature of $500 - 700°C$.

Several studies have been done on low temperature synthesis of rare earth high T_c superconductors. These processes are classified into two types:

(i) deposition at a substrate temperature above the crystallizing temperature $T_{cr}(500 - 700°C)$ followed by postannealing at a lower temperature $(400 - 600°C)$ in O_2.

(ii). deposition at a substrate temperature above the crystallizing temperature without any additional postannealing process.

Type (i) corresponds to process (2). In type (i) post annealing is conducted successively in the deposition equipment without breaking the vacuum. In type (ii), post annealing proceeds in the oxygen furnace after deposition. The former (type i) is called, "in situ annealing" and the latter (type ii) "ex situ annealing". Type (ii) corresponds to process (3) which is called "in situ deposition".

The low temperature process with "in situ annealing" or "ex situ annealing" was studied in several deposition processes including sputtering (178), pulsed laser deposition (179), and reactive deposition (180). In these processes, the temperature of the substrates during deposition is $500 - 700°C$. The annealing is done at around

$400 - 500°C$ in O_2. However, the low temperature process without annealing, in situ deposition, is readily available for making thin film electronic devices since it achieves the formation of the multi-layered structure of the high T_c superconductors.

In the YBC system if the substrate temperature during deposition T_s satisfies the following relationship

$$T_e \leq T_s \leq T_t, \tag{5.12}$$

where T_e denotes the epitaxial temperature, and T_t the transition temperature from the tetragonal to the orthorhombic phase, and enough oxygen is supplied onto the film surface during deposition so as to oxidize the deposited films. The as-deposited films will show the single crystal phase and exhibit superconductivity without the postannealing process.

In situ deposition of the YBC system has been attempted by magnetron sputtering. It has also been confirmed that irradiation of oxygen ions and/or plasma onto the deposited film during deposition is important in order to achieve in situ deposition. Under suitable irradiation of oxygen plasma onto the film surface, excellent superconducting transition temperature was observed in as deposited Er-Ba-Cu-O thin films: The onset temperature was 95 K with a zero resistance temperature of 86 K for films deposited at 650°C (69).

Several advantages have been found for the in situ deposition such as a smooth surface of the deposited films and small interdiffusion between deposited superconducting films and substrates (152). In situ deposition, however, has exhibited low critical current density due to the presence of crystal boundaries in the deposited films which will form weak-links. The magnitude of critical currents is strongly affected by the application of an external magnetic field when its direction is parallel to the c-axis of the oriented superconducting films (181). These weak points observed in in situ deposition may result from imperfect crystallinity of the deposited films, which is essentially improved by refinement of the deposition system.

In rare earth free high T_c superconductors of the BSCC and/or TBCC systems several different superconducting phases are simultaneously formed during the postannealing process. Although the basic thin film processes for rare earth free high T_c superconductors are essentially the same as those of rare earth high T_c superconductors, the simultaneous growth of the different superconducting phase causes difficulty in the controlled deposition of the single phase high T_c superconducting thin films.

5.1.4.5 Deposition: rare earth high T_c superconductors: The simplest method for making rare earth high T_c films is the deposition of amorphous films by sputtering at a low substrate temperature followed by a postannealing process (process (1)). Typical sputtering conditions for YBC films are shown in Table 5.26. These targets were obtained by reacting a mixed powder of Y_2O_3 (99.99%), $BaCO_3$ (99.99%) and CuO (99.9%) in air at 900°C for 8 hours and then sintering at 900°C for 8 hours in air.

Target	$(Y_{0.4}Ba_{0.6})_3Cu_3O_x$
	(100 mm in dia.)
Substrate	$(1\bar{1}02)$ plane of sapphire
Substrate temperature	200°C
Sputtering gas	Ar
Gas pressure	0.4 Pa
Rf input power	150 W
Growth rate	150 Å/min

Table 5.26: Sputtering conditions for the deposition of high T_c thin films (process (1)).

The measurement of resistivity was carried out using the standard four-probe technique with gold electrodes fabricated on the surface of the films. The measured current density was about 5A/cm². Samples were fixed to the copper block and the temperature was measured by a Chromel-Au (Fe) thermocouple attached to the copper block. Figure 5.93 shows photographs of the samples. The as-sputtered films were insulating with brown color. After postannealing at 900°C for 1 hour in O_2 the films showed superconductivity as shown in Fig. 5.94.

Figure 5.93: A photograph of the thin film superconductors for the measurements of resistive properties.

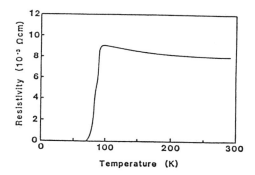

Figure 5.94: Temperature dependence of resistivity for sputtered Y-Ba-Cu-O thin films on (1$\bar{1}$02) sapphire, process (1).

The onset temperature was 94 K with zero-resistivity at 70K (68). These annealed YBC films showed a polycrystalline phase with the preferred orientation of (103) crystal axis perpendicular to the substrate as shown in Fig. 5.95. The (103) surface of YBC corresponds to the closest packed plane. The temperature dependence of resistivity shown in Fig. 5.94 suggests that YBC films are composed of the superconducting orthorhombic phase and the semiconducting tetragonal phase since the temperature dependence above the transition temperature is semiconductive. The relatively high resistivity, 2mΩcm at the transition temperature, may also result from the presence of grain boundaries in the YBC films. The sharp superconducting transition with a single superconducting ortho-I phase is obtained for stoichiometric composition in deposited films. Stoichiometric composition is successfully achieved by multi-layer deposition.

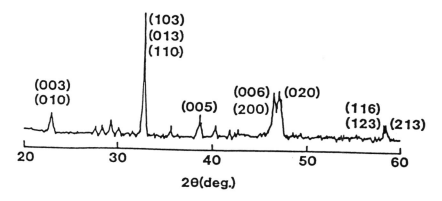

Figure 5.95: Typical X-ray diffraction pattern of sputtered Y-Ba-Cu-O thin film on (100) SrTiO$_3$ (process (1)).

Improvement of the crystallinity of high T$_c$ films is achieved by deposition at a higher substrate temperature (process (2)). Table 5.27 shows typical sputtering conditions for the improvement of crystalline properties of GBC films. Since the concentrations of Ba and Cu in GBC films are reduced at higher substrate temperatures, the composition of the target is modified so as to achieve stoichiometric composition for sputtered GBC films (182). Typical superconducting properties of these GBC films are shown in Fig. 5.96. The low resistivity, less than 0.5mΩcm, at the transition temperature corresponds to bulk resistivity was observed for YBC films. The temperature dependence of resistivity was metallic at temperatures above the transition temperature. Similar properties were also obtained by electron beam deposition (66,183).

Target	$GdBa_2Cu_3O_x$
	(100 mm in dia.)
Substrate	(100) plane of MgO
Substrate temperature	600 and 750°C
Sputtering gas	$Ar + O_2$ (3 : 2)
Gas pressure	0.4 Pa
Rf input power	130 W
Growth rate	80 Å/min
Target-substrate spacing	25 - 35 mm

Table 5.27: Sputtering conditions of low temperature deposition for Gd-Ba-Cu-O thin films, process (2).

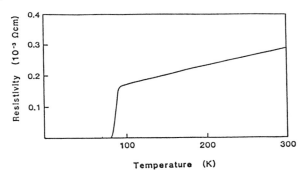

Figure 5.96: Temperature dependence of resistivity for sputtered Gd-Ba-Cu-O thin films on (100) MgO, process (2).

However, epitaxial YBC thin films with single phase YBC are generally difficult to deposit, since the composition of the films often differ from the stoichiometric value of YBC at substrate temperatures above $T_{cr}(T_{cr} \approx 500 - 700°C)$. In contrast, epitaxial LSC thin films are easily deposited since LSC is composed of a solid solution.

Typical sputtering conditions are shown in Table 5.28 (184). The target was stoichiometric $(La_{0.9}Sr_{0.1})_2CuO_4$ and was made by sintering a mixture of La_2O_3 (99.99%), $SrCO_3$ (99.9%), and CuO (99.9%) at 900°C in air for about 8 hr. The as-sputtered films were conductive with a black color similar to the target. Electron-probe X-ray micro-analyses showed that the concentration of La, Sr, and Cu was close to the target composition. The electron diffraction pattern suggested that an excellent single crystal was epitaxially grown on the substrate. However, when LSC films were deposited on (100)

MgO, the resultant films showed the polycrystalline phase. This is due possibly to the large lattice mismatch between LSC and MgO.

Target	$(La_{0.9}Sr_{0.1})_2CuO_4$
	(100 mm in dia.)
Substrate	(100) plane of $SrTiO_3$
Substrate temperature	600 °C
Sputtering gas	Ar
Gas pressure	0.4 Pa
Rf input power	150 W
Growth rate	100 Å/min

Table 5.28: Sputtering conditions, process (2).

These as-sputtered films showed semiconductive behavior. The superconductivity was observed after postannealing in air at 900°C for 3 days. Typical electron diffraction patterns and electrical properties of these sputtered films are shown in Fig. 5.97. Single crystal LSC films grown on (100) $SrTiO_3$ exhibited excellent superconducting properties. The onset temperature was $\simeq$ 34 K with T_c = 25K. The narrow transition width less than 3 K suggests that these sputtered films are composed of the single phase of layered perovskites K_2NiF_4.

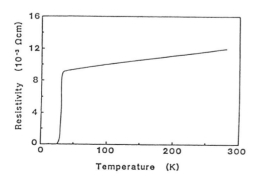

Figure 5.97: Electron diffraction pattern and temperature dependence of resistivity for sputtered La-Sr-Cu-O thin films on (100) $SrTiO_3$, process (2).

However, these processes still need the troublesome postannealing process. Postannealing induced diffusion at the film and substrate interface was conducted at annealing temperatures above $800 - 900°C$. This causes a broad transition due to mutual diffusion between substrate and the deposited films (153).

A discharge plasma of oxygen in sputtering is considered suitable for oxidation of thin films during the deposition. Thus, if the thin films are immersed in the oxygen plasma during deposition in situ deposition (process (3)) will potentially be achieved (185).

Sputter deposition with two target-substrate spacings, 35 mm and 40 mm, was used in the preparation of Er-Ba-Cu-O films on MgO. Typical sputtering conditions are shown in Table 5.29. The target was made by sintering the mixture of Er_2O_3, $BaCO_3$ and CuO at 900°C for 20 hr in air. The surface of the substrate is exposed to discharge plasma with a target spacing of 35 m. For spacing of 40mm the substrate is situated outside of the plasma.

Target	$Er_1Ba_2Cu_{4.5}O$
	(100 mm in dia.)
Substrate	$(100)MgO$ and $(110)SrTiO_3$
Sputtering gas	$Ar + O_2$ 4 : 1
Gas pressure	0.4 Pa
Rf input power	175 W
Substrate temperature	650°C
Growth rate	70 Å/min

Table 5.29: Sputtering conditions, process (3).

The temperature dependence of resistivity for as-sputtered films is shown in Fig. 5.98. The film made with spacing of 35 mm showed a sharp superconducting transition with onset at 92 K and $T_{R=0} = 86K$. On the other hand, the film with spacing of 40 mm exhibited a much broader superconducting transition and zero resistance was realized at 57 K. It is considered that the effect of target spacing on superconducting properties results from the difference of oxidation in the films. We can only roughly presume oxidation of the films from the crystalline information obtained. Sufficient oxidation leads surely to the superconducting orthorhombic structure, while oxygen defects cause the semiconducting tetragonal structure.

Figures 5.99 (a) and 5.99 (b)show the x-ray diffraction patterns of films made with spacings of 35 mm and 40 mm, respectively. The c-axis is primarily oriented perpendicular to the film plane. The crystal system can be discriminated by the lattice constant c, i.e., c = 11.68 Å for the orthorhombic structure (O) and c = 11.8 - 11.9 Å for the tetragonal structure (T). The film made with spacing of 35 mm shows a mixed structure with dominant orthorhombic and minor tetragonal phases. On the other hand, the film with spacing of 40 mm shows the tetragonal structure. From these results it is evident that oxidation progresses more for a spacing of 35 mm than for 40 mm. For comparison, the X-ray diffraction pattern of the film $T_{R=0} = 55K$ made with a target-to-sample distance of 35mm and at a higher substrate temperature of 700°C is shown in Fig. 5.99(c). The film shows the tetragonal structure. Since deposition was carried out at a temperature higher than the T-O transition, the oxidation in passing through the T-O point was not sufficient for quick cooling.

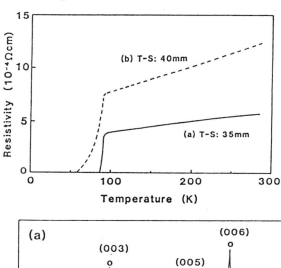

Figure 5.98: Temperature dependence of the resistivity for Er-Ba-Cu-O thin films on (100) MgO, process (3).

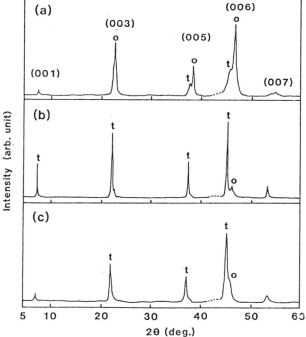

Figure 5.99: X-ray diffraction patterns for Er-Ba-Cu-O thin films on (100) MgO, process (3).

On SrTiO₃ (110) substrates, epitaxial films are prepared by the same process. Figure 5.100 shows the RHEED pattern of the epitaxial Er-Ba-Cu-O film. The temperature dependence of resistivity for as-deposited films shows similar characteristics and zero-reactivity is realized below 80 K. Y-Ba-Cu-O films are also prepared by this process.

Figure 5.100: Electron diffraction pattern of the epitaxial Er-Ba-Cu-O thin film on (100) SrTiO₃, process (3).

These facts suggest the possibility of in-situ deposition (process (3)) in Table 5.25, although the in-situ deposited films are not composed of the single phase of the orthorhombic structure.

The experiments on Gd-Ba-Cu-O thin films suggest that in situ postannealing in O_2 at the relatively low temperature of $400 - 600°C$ increases the orthorhombic phase and improves the superconducting properties (186). The effects of low temperature postannealing are also verified in pulsed laser deposition.

5.1.4.6 Deposition: rare earth free high T_c superconductors: In the rare earth free high T_c superconductors several superconducting phases are present for different chemical compositions. Typical chemical compositions for the Bi-system and Tl-system are listed in Table 5.30. Their superconducting properties have not yet been fully explained.

			Tc (k)	Institute	Date
Bi-system : $Bi_2O_2 \cdot 2SrO \cdot (n-1)Ca \cdot nCuO_2$					
	$Bi_2Sr_2CuO_6$	(2 2 0 1)	7 ~ 22	Caen Univ. (France) Aoyamagakuin Univ. (Japan)	1987.5
	$Bi_2Sr_2CaCu_2O_8$	(2 2 1 2)	80	National Res. Institute for Metals (Japan)	1988.1
	$Bi_2Sr_2Ca_2Cu_3O_{10}$	(2 2 2 3)	110	National Res. Institute for Metals (Japan)	1988.3
	$Bi_2Sr_2Ca_3Cu_4O_{12}$	(2 2 3 4)	~ 90	Matsushita Elec. (Japan)	1988.9
Tl-system : $Tl_2O_2 \cdot 2BaO \cdot (n-1)Ca \cdot nCuO_2$					
	$Tl_2Ba_2CuO_6$	(2 2 0 1)	20 ~ 90	Institute for Molecular Sci. (Japan) Arkansas Univ. (U.S.A.)	1987.12
	$Tl_2Ba_2CaCu_2O_8$	(2 2 1 2)	105	Arkansas Univ. (U.S.A.)	1988.2
	$Tl_2Ba_2Ca_2Cu_3O_{10}$	(2 2 2 3)	125	Arkansas Univ. IBM (U.S.A.)	1988.3
: $TlO \cdot 2BaO \cdot (n-1)Ca \cdot nCuO_2$					
	$TlBa_2CaCu_2O_7$	(1 2 1 2)	70 ~ 80	IBM (U.S.A.)	1988.5
	$TlBa_2Ca_2Cu_3O_9$	(1 2 2 3)	110 ~ 116	IBM (U.S.A.)	1988.3
	$TlBa_2Ca_3Cu_4O_{11}$	(1 2 3 4)	120	ETL (Japan)	1988.5
	$TlBa_2Ca_4Cu_5O_{13}$	(1 2 4 5)	< 120	ETL (Japan)	1988.5

Table 5.30: Rare-earth-free high T_c superconductors.

Thin films of Bi-Sr-Ca-Cu-O system are prepared by rf-planar magnetron sputtering similar to YBC films. The target is complex oxides of Bi-Sr-Ca-Cu-O which is made by sintering a mixture of Bi_2O_3 (99.999%), $SrCO_3$ (99.9%), $CaCO_3$(99%) and CuO (99.9%) at 880°C for 8 hr in air.

It is known that superconducting properties are strongly affected by the substrate temperature during deposition. Figure 5.101 shows typical X-ray diffraction patterns with resistivity-temperature characteristics for Bi-Sr-Ca-Cu-O thin films about 0.4μm thick deposited at various substrate temperatures. It shows that films deposited at 200°C exhibit a $Bi_2Sr_2CaCu_2O_x$ structure with the lattice constant c = 30.64Å which corresponds to the low T_c phase (187). The films show zero resistance temperature of $\simeq$70K Fig. 5.101 (a).

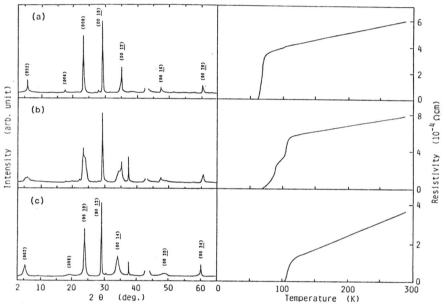

Figure 5.101: X-ray diffraction patterns with resistivity versus temperature for the annealed Bi-Sr-Ca-Cu-O films: substrate temperature during deposition; (a) 200°C, (b) 700°C, (c) 800°C.

When the substrate temperature is raised during deposition the high T_c phase with $T_c\simeq$ 110K, the $Bi_2Sr_2Ca_2Cu_3O_x$ structure with the lattice constant c$\simeq$36Å, is superposed on the X-ray diffraction pattern (Fig. 5.101 (b))(188). At the substrate temperature of around 800°C a single high T_c phase is observed. The films show zero resistance temperature of 104 K (Fig. 5.101 (c)).

Typical sputtering conditions are shown in Table 5.31. The target is complex oxides of Bi-Sr-Ca-Cu-O. The compositions near the 1-1-1-2 ratio of Bi-Sr-Ca-Cu. Processes (1) and/or (2) are used for deposition. Single crystals of (100)MgO are used as substrates. The superconducting properties are improved by postannealing at 850 − 900°C in 5 hr for O_2 (189).

Target	Bi:Sr:Ca:Cu:=1-1.7:1:1-1.7:2
	(100 mm in dia.)
Sputtering gas	Ar/O$_2$=1-1.5
Gas pressure	0.5 Pa
Rf input power	150 W
Substrate temperature	200 - 800°C
Growth rate	80 Å/min

Table 5.31: Sputtering conditions for Bi-Sr-Ca-Cu-O thin films.

It is noted that the formation of these superconducting phases strongly depends on the annealing temperature and chemical composition as shown in Fig. 5.102 (190).

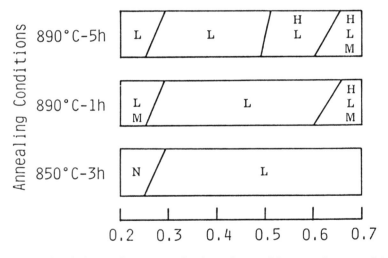

Figure 5.102: Variations of superconducting phase with annealing conditions for Bi$_2$(Sr$_{1-x}$Ca$_x$)$_{n+1}$Cu$_n$O$_y$ thin films: H, high T$_c$ phase (n=3); L, low T$_c$ phase (n=2); M & N, other phases (Satoh (1988) (190)).

Similar to the Bi-Sr-Ca-Cu-O system, thin films of the Tl-Ba- Ca-Cu-O system are prepared by rf-magnetron sputtering on a MgO substrate. Typical sputtering conditions are shown in Table 5.32. However, their chemical composition is quite unstable during deposition and the postannealing process due to the high vapor pressure of Tl. Thin films of the Tl system are deposited without intentional heating of substrates (< 200°C) and are annealed at 890 − 900°C in Tl vapor (80). It is seen that the superconducting phase of the resultant films strongly depends on postannealing conditions.

Target	Tl:Ba:Ca:Cu:=2:1-2:2:3
	(100 mm in dia.)
Sputtering gas	Ar/O$_2$=1
Gas pressure	0.5 Pa
Rf input power	100 W
Substrate temperature	200°C
Growth rate	70 Å/min

Table 5.32: Sputtering conditions for Tl-Ba-Ca-Cu-O thin films.

Figure 5.103 shows typical X-ray diffraction patterns with resistivity-temperature characteristics for Tl-Ba-Ca-Cu-O thin films annealed at different conditions.

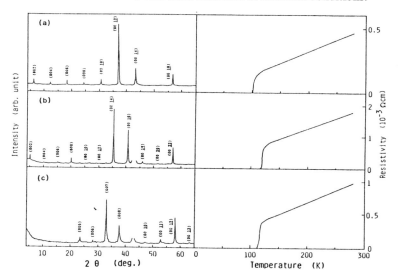

Figure 5.103: X-ray diffraction patterns with resistivity versus temperature for the annealed Ti-Ba-Ca-Cu-O films.

The 0.4μm thick film exhibits the low temperature phase, Tl$_2$Ba$_2$CaCu$_2$O$_x$ structure, with the lattice constant c$\approx$29Å after slight annealing at 1 min at 900°C (Fig. 5.103a). The 2 μm thick films annealed at 900°C 13 min show the high temperature phase, Tl$_2$Ba$_2$Ca$_2$Cu$_3$O$_x$ structure with the lattice constant c$\approx$36Å (Fig. 5.103b). In specific annealing conditions the other superconducting phase TlBa$_2$Ca$_3$Cu$_4$O$_x$ structure with the lattice constant c$\approx$19Å is also obtained (Fig. 5.102c).

5.1.4.7 Structure and Structural Control: As described in a previous section, thin film processing of high T$_c$ superconductors is classified into three processes: deposition at low substrate temperature with postannealing (process (1)), deposition at high temperature with postannealing (process (2)), and deposition at high temperature without postannealing (process (3)).

One of the most important problems to be solved for thin film processing is lowering the synthesis temperature. At present, lowering of the synthesis temperature can be achieved by both process (2) and process (3) for rare-earth high T_c superconductors.

In process (2) lowering of the synthesis temperature can be achieved with low temperature post annealing at around $400 - 600°C$. The minimum synthesis temperature is determined by the crystallizing temperature of the high T_c superconductors in these processes. The crystallizing temperature of YBC, for instance, is around $500 - 700°C$ It is noted that in the process (3) as-deposited films show superconducting properties without any postannealing, i.e. in-situ deposition.

As seen in the X-ray diffraction pattern, the in-situ deposited films are composed of the orthorhombic phase and the tetragonal phase. The TEM image suggests that these films are composed of small crystallites as shown in Fig. 5.104. The TEM image of the sputtered films also denotes the presence of crystal boundaries as shown in Fig. 5.105. This may reduce the critical current J_c The J_c for in-situ deposited Er-Ba-Cu-O thin films is proportional to $(1 - T/T_c)^{1.8}$, which is close to $(1 - T/T_c)^{1.5}$. This indicates that the current transport will be partially governed by the weak link of superconductive regions. At present high critical current is obtained in process (2) using postannealing (190).

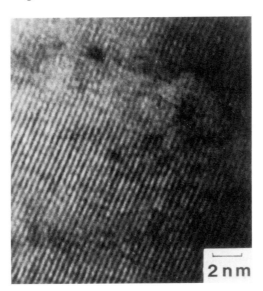

Figure 5.104: TEM image of in situ deposited Er-Ba-Cu-O thin films on (100) MgO.

MgO (100)

Although low temperature synthesized films are not perfect single crystals, the low temperature process gives several favorable properties such as the suppression of interdiffusion at the film and substrate interface as shown in Fig. 5.106.

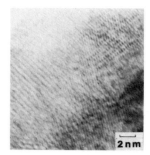

SrTiO₃ (110)

Figure 5.105: TEM image of in situ deposited Er-Ba-Cu-O thin films on (110) SrTiO₃.

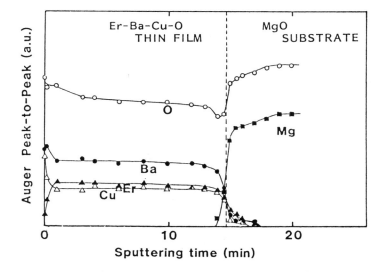

Figure 5.106: Auger depth profile of Er-Ba-Cu-O thin films of 2000 Å thick deposited on (100) MgO substrate by the low temperature process without the postannealing.

As described in the previous section for YBC high T_c superconductors, the superconducting orthorhombic phase is stabilized during the postannealing process. For rare earth free high T_c superconductors of the Bi-system, the superconducting phases of the sputtered films are controlled by the substrate temperature during deposition; the low T_c phase of $Bi_2Sr_2CaCu_2O_x$ system is obtained at the substrate temperature below 600°C, and the high T_c phase of the $Bi_2Sr_2Ca_2Cu_3O_x$ system is obtained at the substrate temperature above 750°C.

The SEM image suggests that the thin films of the Bi-system are composed of mica-like crystallites as shown in Fig. 5.107. The c-axis of the crystallites is perpendicular to the crystal plane. The large crystallites allow for the large critical current.

The critical current density measured for the Bi-Sr-Ca-Cu-O films is as high as $2 \times 10^5 A/cm^2$ at 77 K and $6 \times 10^6 A/cm^2$ at 4.2 K. The critical current density at 77 K will be governed by the high T_c phase. The current will flow through the current channel presented in the sputtered Bi-Sr-Ca-Cu-O films since the films are composed of a mixture of the high T_c phase and low T_c phase. A higher critical current density will be possible in the case of films with a single high T_c phase. The diamagnetic measurement suggests that the film is composed of 5-10% of the high T_c phase (191). This suggests that the net critical current of the high T_c phase will be $2 \times 10^6 \simeq 4 \times 10^6 A/cm^2$ at 77 K and $6 \times 10^7 \simeq 1.2 \times 10^8 A/cm^2$ at 4.2 K.

Bi-Sr-Ca-Cu-O film

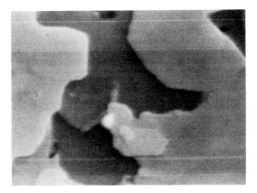

Figure 5.107: SEM image of Bi-Sr-Ca-Cu-O thin films with c-axis orientation.

1μm

It is noted that the temperature variations of the J_c are governed by $(1 - T/T_c)^2$ (192). The square power dependence is different from the 3/2 power dependence predicted by the well-studied proximity junction tunneling model, which is based on the BCS theory. The presence of the layered structure will cause the square power dependence. Similar properties are observed in the Tl-Ba-Ca-Cu-O films (193).

In the crystallites of low T_c Bi-Sr-Ca-Cu 2-2-1-2 phase, the atomic arrangements are found to be uniform as indicated in the TEM image shown in Fig. 5.108. However, in the crystallites of the high T_c Bi-Sr-Ca-Cu 2-2-2-3 phase, the crystallites are composed of the different superconducting phases including Ba-Sr-Ca-Cu 2-2-1-2, 2-2-3-4, and 2-2-4-5 phases, although the resistivity-temperature characteristics correspond to the single superconducting phase of the 2-2-2-3 structure. The presence of the mixed phase is also confirmed by the spreading skirt observed in the X-ray diffraction pattern at the low angle peak around $2\Theta \simeq 4°$.

It is reasonable to consider that the presence of the mixed phases results from the specific growth process of the present rare earth free superconducting thin films: The rare earth free superconducting thin films may be molten during the annealing process and the superconducting phase will be formed during the cooling cycle (194).

For the Tl-Ba-Ca-Cu-O system, the sputtered films exhibit rough surface morphology as shown in Fig. 5.109. The critical current is lower than that of the Bi-system.

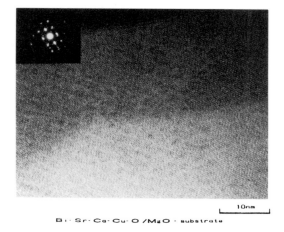

B i - Sr - Ca - Cu - O /MgO · substrate

Figure 5.108: TEM image of the sputtered Bi-Sr-Ca-Cu-O thin films of 2-2-1-2 structure.

Tl-Ba-Ca-Cu-O film

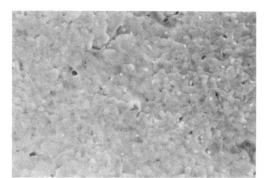

Figure 5.109: SEM image
of Tl-Ba-Ca-Cu-O thin films.

10μm

Figure 5.110 shows XPS measurements for the crystallized Bi-Sr-Ca-Cu-O films. It
shows that the annealing process modifies the crystal structure near the $Cu - O_2$ layer
and increases the density of Cu^{3+}. It is also noted that the Sr 3d and/or Ca 2p electron
spectra move during the annealing. This implies the some structural changes will appear
around Sr and/or Ca sites during the annealing. In contrast, the Bi-O layered structure is
stable during annealing (195). This implies that the single superconducting phase will be
synthesized when the Bi-O basic structure is crystallized and the stoichiometric composition is kept for the unit cell of Bi-Sr-Ca-Cu-O.

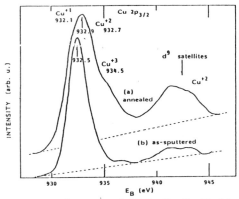

Figure 5.110: XPS spectrum of sputtered Bi-Sr-Ca-Cu-O thin films: (a) post-annealed in O_2 at 845°C, 300 min, (b) as sputtered.

5.1.4.8 Phase Control by Layer–by–layer Deposition: The present high T_c supercon-ductors are composed of layered oxides. If the layered oxides are atomically synthesized by layer-by-layer deposition the superconducting phase will be closely controlled. These considerations have been successfully confirmed by sputter deposition of the Bi-systems in the multi-target sputtering process shown in Figure 5.111. The deposition rate

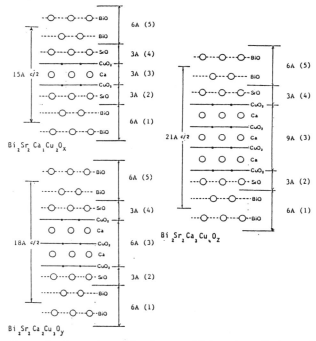

Figure 5.111: Layer-by-layer deposition by a multi-target sputtering: alternative depo-sition in the order (1)->(2)->(3)->(4)->(5).

is selected so as to deposit the Bi-O, Sr-O, Cu-O layers in an atomic scale range. The substrate temperature was kept around the crystallizing temperature of 650°C. X-ray diffraction analyses suggest that as-sputtered films show the Bi layered oxide structure with broad superconducting transitions. These superconducting properties were improved by postannealing at $850 - 900°C$ in O_2. Figure 5.112 shows typical results for layer-by-layer deposition with postannealing. It is noted that phase control is achieved simply by the amounts of Cu-Ca-O during the layer-by-layer deposition. Experiments show that the T_c does not increase monotonously with the number of the Cu-O layers. In the Bi-layer system, the T_c shows the maximum, 110K at, three layers of Cu-O, $Bi_2Sr_2Ca_2Cu_3O_x$. At four layers of Cu-O, $Bi_2Sr_2Ca_3Cu_4O_x$, the T_c becomes 90K (196).

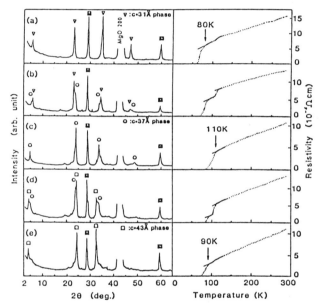

Figure 5.112: X-ray diffraction patterns with resistive-temperature characteristics for the phase-controlled Bi-Sr-Ca-Cu-O thin films: (a) $Bi_2Sr_2CaCu_2O_x$, (b) $Bi_2Sr_2CaCu_2O_x$ / $Bi_2Sr_2CaCu_3O_y$, (c) $Bi_2Sr_2Ca_2Cu_3O_y$, (d)$Bi_2Sr_2Ca_2Cu_3O_y$/ $Bi_2Sr_2Ca_3Cu_4O_z$ and (e) $Bi_2Sr_2Ca_3Cu_4O_z$.

The layer-by-layer deposition is one of the most promising processes for fine control of the superconducting phase. Improvements of crystallinity during the layer-by-layer deposition will reduce the annealing temperature and allow the low temperature process and/or in situ deposition. Artificially-made layered oxide superconductors (ALOS) can be also synthesized by layer-by-layer deposition (196).

5.1.4.9. Diamagnetization Properties: Magnetization properties are measured by an rf SQUID susceptometer. The operation conditions of an rf SQUID are listed in Table 5.33. The diamagnetization of the high T_c oxide superconducting films with c-axis orientation essentially indicates the anisotropy, when the external field is applied both perpendicularly and parallel to the c axis. The high diamagnetization is observed when the film plane is perpendicular to the magnetic field.

Range of measurements	± 2 emu
Accuracy	
magnetic flux	1×10^{-8} emu/$\sqrt{Hz}$
susceptance	1×10^{-10} emu/cm^3/$\sqrt{Hz}$
Applied magnetic field	± 10 kGauss
Sample dimension	5 mm × 5 mm in dia.

Table 5.33: Operating conditions of rf-SQUID susceptor for the measurements of diamagnetization properties.

Typical diamagnetic hysteresis loops of superconducting oxide films measured at 4.2 K are shown in Figure 5.113 (197). These loops show the so-called "Lenz law" (198). When the external field is decreased to a small extent while holding the same field direction, diamagnetization is reversed to the opposite direction within a very short time, keeping its absolute value. Similar hysteresis loops are also observed in the single crystals (199). It is noted that small additions of sulfur into the YBC system increases diamagnetization (176).

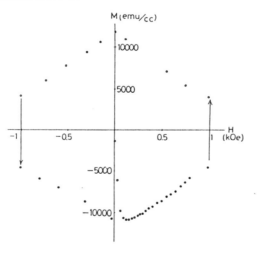

Figure 5.113: Magnetic hysteresis curve for Bi-Sr-Ca-Cu-O thin films measured at 4.2 K.

Temperature variations of the diamagnetization of c-axis oriented superconducting films are measured both for cooling the specimen in the external field (Meissner effect), and for the warming specimen by applying the field after zero-field cooling (shield effect). Typical results for Bi-Sr-Ca-Cu-O thin films are shown in Fig. 5.114.

By using Bean's formula the critical current $J_c(A/cm^2)$ becomes

$$J_c = 30M/\gamma,$$

(5.13)

where M denotes the diamagnetization (emu/cm^3), γ (cm) the effective radius of the sample specimen.   For Bi-Sr-Ca-Cu-O thin films with $\gamma = 0.1cm$, J_c becomes $3.3 \times 10^6 A/cm^2$ at 4.2 K, the value found to be very close to the J_c measured by transport measurements (198).

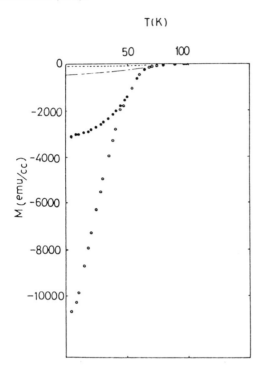

Figure 5.114: The temperature dependence of diamagnetization for Bi-Sr-Ca-Cu-O film; open circles: shielding effect in 140 Oe; closed circles: shielding effect in 10 Oe; dash-dot: Miessner effect in 140 Oe; dotted line: Meissner effect in 10 Oe.

5.1.4.10 Passivation of Sputtered High T_c Thin Films: It is of practical importance to form passivation films on superconducting films in order to reduce environmental influences such as humidity. For such purposes, inorganic insulating films are considered more preferable than organic materials. Aluminum oxide films were formed on Y-Ba-Cu-O thin films with sapphire substrates by rf-magnetron sputtering, and reportedly Ba atoms were incorporated in the Al_2O_3 films. In this case the substrates were exposed to $Ar - O_2$ plasma. It is considered more desirable to form passivation films without exposing superconducting films directly to the plasma discharge. Film preparation should be performed at the lowest temperatures possible in order to minimize the influence on crystal structures. One promising method for low temperature deposition is the reactive evaporation method (REM) (163).

Figure 5.115 shows a schematic configuration of the system for electron-cycloton-resonance (ECR) REM. It is composed of an ECR plasma source, electron evaporation source (e-gun) and a vacuum chamber where substrates are to be located. The microwave frequency is 2.45 GHz and the magnetic flux density is 875 Gauss. Silicon is evaporated using an electron beam gun. For Si-N and Si-O film depositions, N_2 and O_2 gas are introduced, respectively. Si evaporation and ECR plasma irradiations are simultaneously performed for Si-N and Si-O formations. Deposition conditions are shown in Table 5.34.

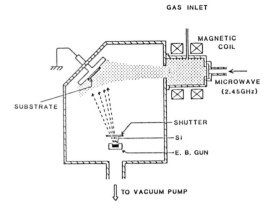

GAS INLET

MAGNETIC COIL

MICROWAVE (2.45GHz)

SUBSTRATE

SHUTTER

Si

E. B. GUN

TO VACUUM PUMP

Figure 5.115: A schematic configuration of the plasma assisted electron beam deposition. The oxygen plasma is supplied by the ECR plasma source.

Passivation film	SiN	SiO
Gas flow rate	N_2 50 sccm	O_2 20 sccm
Gas pressure	1×10^{-3} Torr	8×10^{-4} Torr
Microwave power	600 W	600 W
Deposition rate (Si)	1-5 A/s	1 A/s

Table 5.34: Typical deposition conditions of Si-N, Si-O passivation films.

Figure 5.116 represents the depth distributions of the compositional elements in the Si-O/Gd-Ba-Cu-O system as determined by AES measurement. It is seen that the compositional elements of Gd-Ba-Cu-O films are not detected in the Si-O film, although the depth distributions of both compositional elements in the Si-O/Gd-Ba-Cu-O system around the interface are slightly complicated. The depth distribution of Si around the interface is not symmetrical in both sides of the Si-O and Gd-Ba-Cu-O films. At the depth where the Si signal is not detected, the signals caused by Gd, Ba and Cu still increase with depth. It is deduced that the resultant depth distributions are caused by the film coverage over the superconducting Gd-Ba-Cu-O film when the superconducting film surface is not smooth, since the Si-O film formation is essentially carried out obliquely to the Gd-Ba-Cu-O film. Figure 5.117 shows the temperature dependences of resistivities for the Gd-Ba-Cu-O film before and after Si-O film deposition. The onset temperature of the as-deposited Gd-Ba-Cu-O is 88 K and the zero-point $(T_{R=0})$ is 57 K. After Si-O film formation, the $T_{R=0}$ increased slightly to 62 K. This may be caused by oxygen ECR plasma exposure to the sample surface during the initial stage of Si-O film deposition.

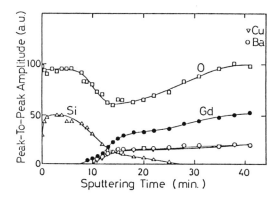

Figure 5.116: Depth distribution of the compositional elements in Si-O/Gd-Ba-Cu-O system determined by AES measurement.

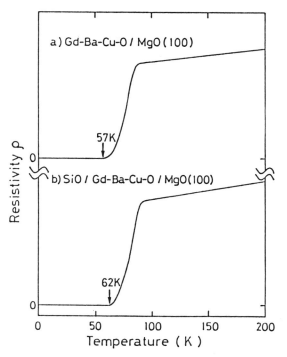

Figure 5.117: Superconducting properties of sputtered Gd-Ba-Cu-O thin films deposited on (100) MgO substrates with and without overcoating of a Si-O layer.

In the case of Si-N film formation on the superconducting thin film of Gd-Ba-Cu-O, both the values of T_c and $T_{R=0}$ after Si-N film deposition are the same as those of the as-deposited Gd-Ba-Cu-O film as shown in Table 5.35.

Hall measurements suggest that the high T_c superconductors, LSC, YBC, BSCC, and TBCC show p-type conduction. The postannealing of the used for achieving superconducting properties will act as a hole injection process. In contrast, the Nd-Ce-Cu-O systems are known as n-type superconductors. The postannealing is done at reducing atmosphere. It is noted the density of Cu^{+1} will increase during the postannealing process (200).

	Tc(K)	$T_{R=0}$ (K)
Virgin Gd-Ba-Cu-O thin film	88	57
SiO/Gd-Ba-Cu-O thin film	89	62
SiN/Gd-Ba-Cu-O thin film	89	58

Table 5.35: Influence of overcoating of dielectric layers on superconducting properties for Gd-Ba-Cu-O thin films.

Table 5.36 shows the physical properties of present high T_c superconducting films.

	$La_{1-x}Sr_xCuO_4$ x=0.05	$YBa_2Cu_3O_{7-x}$	Bi-Sr-Ca-Cu-O (2212)	(2223)	Tl-Ba-Ca-Cu-O (2212)	(2223)
$Tc_{R=0}$(K)	30	84	80	104	102	117
-dHc2$_{\parallel}$/dT (T/K)	6.3	4.6	7.3	7	20	-
-dHc2$_\perp$/dT (T/K)	1.2	0.54	0.36	0.34	0.6	-
Hc2$_{\parallel}$(0) (T)	130	373	440	500	1408	-
Hc2$_\perp$(0) (T)	25	44	20	24	42	-
$\xi_{\parallel}$ (A)	37	27	12.8	11.7	28	-
$\xi_\perp$ (A)	7.1	3.2	2.7	2.6	0.8	-
Anisotropy	5.3	8.4	20	20	31	-

Table 5.36: Typical superconducting properties of high T_c superconducting thin films.

5.1.5 Transparent Conducting Films

Thin films of SnO_2 and In_2O_3 are transparent with high electrical conductivity. These conduction films are prepared by a chemical deposition process which includes spray coating or a physical deposition process, a sputtering process, and reactive vacuum evaporation (201,202).

Of these processes, the sputtering process gives the most controlled deposition for this type of conducting film. Generally the conducting films are prepared by dc sputtering from the alloy target of In-Sn in an oxygen atmosphere, or rf-sputtering from the compound target of In-Sn oxides. In dc-sputtering the target surface variation changes the electrical conductivity. In order to achieve high reproducibility of the film properties, the partial pressure of oxygen during sputtering deposition should be closely controlled. The as sputtered films are often annealed in air at 400 to 500°C to increase their transparency

(203). In contrast to dc-sputtering, sputtering of the oxide target produces the conductive transparent films without any post-annealing process.

Table 5.37 shows typical sputtering conditions for the deposition of transparent conductive films. The target is a ceramic of indium tin oxides, In_2O_3 with 5 to 10% SnO_2. The rf-magnetron sputtering, where the working pressure is as low as 1 mTorr , allows the deposition of high conductive films at low substrate temperature. The addition of oxygen of 10^{-4} to 10^{-5} Torr during sputtering increases the crystallinity of the sputtered films and increases the conductivity (204). Since these sputtered films exhibit high transparency in the visual region and high reflectance in the infrared region, they are used for both liquid crystal display and selective coating in resistively heated solar energy conversion systems.

SPUTTERING SYSTEM	RF PLANAR MAGNETRON
TARGET	SINTERED ITO (100 mm in diameter)
SUBSTRATE	FUSED QUARTZ
SEPARATION (target-substrate)	30 mm
SPUTTER GAS	4×10^{-3} Torr(Ar)
SUBSTRATE TEMPERATURE	$40^{\circ}C$
SPUTTER POWER	200 W
SPUTTER TIME	5 - 10 min
AREA RESISTIVITY	10 -100 $\Omega/\square$

Table 5.37: Sputtering conditions for the deposition of transparent conducting films.

It is known that the resistivity of these films is around $10^{-3}\Omega cm$. Lowering of the deposition temperature is required for many applications. Recent experiments suggest that addition of H_2O vapor of 10^{-4} to 10^{-5} Torr during deposition reduces resistivity below $5 \times 10^{-4}\Omega cm$ even if the substrate temperature is lower than $200^{\circ}C$.

5.2 NITRIDES

Most nitrides can be characterized as high temperature materials which show high mechanical strength. A wide variety of electronic properties, from superconductors to dielectrics can be found in various nitrides. McLean and his co-workers performed pioneering works on sputter deposition of TaN films for making highly precise thin film resistors (205). They were used in touch-tone telephones at that time.

Nitride thin films are easily prepared by sputtering, since the vapor pressure of nitrides is generally so low that composition in sputtered films will scarcely shift due to evaporation of one species. A sintered nitride target is used for sputtering in Ar gas, and a metal target for sputtering in a nitride forming atmosphere.

5.2.1 TiN Thin Films

Titanium nitride, TiN, shows a cubic structure of the NaCl type. Thin films of TiN are prepared by sputtering from a TiN powder target in Ar. Table 5.38 shows typical sputtering conditions for deposition of TiN thin films. These sputtered films show a crystalline structure even at a low substrate temperature.

Sputtering conditions		Film properties
Sputter system	Rf-magnetron	
Target .	TiN sintered powder [*]	Polycrystal (cubic)
Sputter gas	4×10^{-2} Torr (Ar 6N)	
Substrate	Fused quartz	(111) orientation
Sputter power	400 W	d= 4.24- 4.25A
Substrate temp.	500°C	$\rho = 2 \times 10^{-4} \Omega$ cm
Growth rate	1.5 m/hr	

* Stainless target dish 100 mm in diameter is used for the powder target materials.

Table 5.38: Sputtering conditions for the deposition of TiN thin films.

5.2.2 Compound Nitride Thin Films

Thin films of compounds Ti-Al-N have high mechanical strength and show a wide range of electrical resistivity. Thin films of Ti-Al-N are prepared by sputtering from the composite target Ti/Al in a mixed gas of Ar and N (206). The films are also prepared by direct sputtering of a mixed powder of TiN and AlN in Ar. Table 5.39 shows the sputtering conditions for the deposition of Ti-Al-N films (207). Figure 5.118 show the resistive properties of sputtered Ti-Al-N films for various compositions. The TiN films have a resistivity of $150\mu\Omega$cm, and a temperature coefficient of resistivity $T_{CR} \approx 300$ppm/$^{\circ}$C; the AlN films, $\rho \approx 2000\mu\Omega$cm, $T_{CR} \approx -400$ppm/$^{\circ}$C. It is noted that when AlN/TiN $\simeq 1.0$, the Ti-Al-N films give zero T_{CR}. For $T_{CR} < +100$ppm/$^{\circ}$C ρ becomes $1800\mu\Omega$cm which is one order higher than thin films of $\beta - $Ta($\approx 180\mu\Omega$cm) or $Ta_2N(\approx 290\mu\Omega$cm).

The Ti-Al-N films are composed of crystalline TiN with amorphous AlN when AlN/TiN $\simeq 1$. Thin films of Ta-Al-N, Ti-Si-N, Ta-Si-N also show similar electrical properties as Ti-Al-N films. Figure 5.119 shows the resistive properties of the ternary compound Ti-Zr-Al-N thin films (208). The ternary composition expands the range of resistivity with small T_{CR} These nitride films are useful for making precise thin film resistors and thin film heaters for making a thermal printer head with high stability.

Sputter system	Rf diode
Target	TiN-AlN mixed powder (2N) 100 mm in diameter stainless dish
Sputter gas	$1.5-5 \times 10^{-2}$ Torr (Ar 5N)
Substrate	glass, alumina
Sputter power	300-400 W
Substrate temperature	$150-500^{\circ}C$
Growth rate	$0.6-1.2 \mu m/hr$

Table 5.39: Sputtering conditions for the deposition of Ti-Al-N thin films.

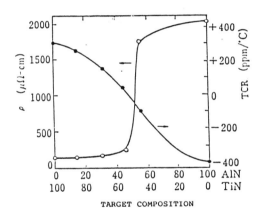

TARGET COMPOSITION

Figure 5.118: Electrical properties of sputtered Ti-Al-N films for various target compositions.

Ti-Zr-Al-N thin film

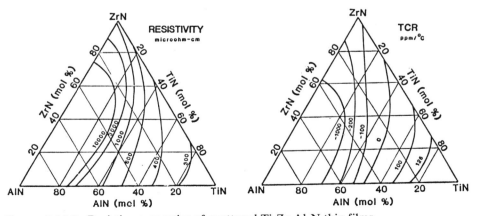

Figure 5.119: Resistive properties of sputtered Ti-Zr-Al-N thin films.

5.2.3 SiN Thin Films

Amorphous films of plasma chemical vapor deposited (CVD) Si-N are produced by the reaction of nitrogen and/or ammonia with silane. These films include hydrogen in the form of N-H and/or Si-H bonds (209). Reduction of the hydrogen concentration in SiN films is necessary for increasing their thermal and chemical stability (210). It is noted that SiN films with low hydrogen concentration can be prepared by ion beam sputter deposition (211).

Figure 5.120 shows the construction of an ion beam sputter system. A mixed gas of argon and nitrogen is introduced into a Kaufman-type ion source. A water cooled Si target is reactively sputtered by the ion beam. The acceleration voltage and ion beam current are 1200 V and 60 mA, respectively.

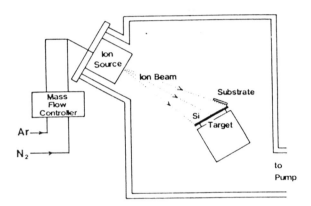

Figure 5.120: Ion-beam sputtering system. Substrate surfaces are irradiated by the ion beam during the deposition. Substrates are located nearly parallel to the ion beam.

The temperature of the substrates, n-type Si (100) wafers, is controlled from room temperature to 200°C. The substrates are located almost parallel to the direction of the ion beam so that the incident ion beam grazes the surface of the substrate during film growth. The vacuum chamber is maintained at about 10^{-4} Torr during sputtering.

Under these conditions transparent, amorphous Si-N films are deposited at a deposition rate of 70Å/ min. Infrared absorption spectra suggest that the sputtered films exhibit a Si-N absorption at about 800 cm^{-1} but absorption due to hydrogen bonds is barely detected. The hydrogen concentration measured, using secondary ion mass spectrometry (SIMS), is found to be below 0.1% which is much lower than that of plasma CVD Si-N films (212).

Table 5.40 shows a summary of the physical properties of sputtered Si-N films and the sputtering conditions. It shows that hydrogen-free Si-N films are prepared by ion-beam sputtering at room temperature. Electron energy loss spectroscopy measurements suggest that the chemical composition of the sputtered films is close to stoichiometric Si_3N_4 when the mixed gas ratio $N_2/Ar > 4$. These sputtered films show high chemical/thermal stability similar to pyrolytic Si_3N_4 films.

Sputtering conditions	
Target	Si (6N, 100 mm in diameter)
Sputter gas	N_2, mixed gas N_2/Ar ($N_2/Ar = 2$–6)
Acceleration voltage	1200 V
Ion beam	60 mA (25 mm in diameter)
Substrate	n-Si (100)
Substrate temp. T_s	RT ~ 200 °C
Deposition rate	70 Å/min
Film thickness	0.1–10 μm
Refractive index (at 6328 Å)	2.1–2.2 ($N_2/Ar > 4$)[a]
Etching rate (buffered HF at 20 °C)	< 30 Å/min ($N_2/Ar > 4$)
Memory trap density	6×10^{11} cm^{-2} (T_s = RT, pure N_2)
in MNOS structure	1×10^{12} cm^{-2} (T_s = RT, $N_2/Ar = 4$)
	6×10^{10} cm^{-2} (T_s = 200 °C, postanneal at 400 °C)
Permittivity	6–7
Dielectric strength	$> 10^6$ V/cm

[a] When $N_2/Ar < 4$, the sputtered films comprise Si-rich SiN showing high refractive index.

Table 5.40: Sputtering conditions and physical properties of sputtered Si-N films.

5.3 CARBIDES AND SILICIDES

Carbides and silicides are known as high temperature materials with strong mechanical strength similar to nitrides. For instance, silicon carbide, SiC, shows a high melting point of 2700°C with Vickers hardness of 4,000kg/mm². The growth of single crystal SiC films has been studied in relation to SiC thin film devices, including high temperature SiC transistors and thin film electro-luminescent devices (213). Diamond thin films are also of technological interest because of their potential applications in electronic devices capable of operating at high temperatures and under irradiation of cosmic rays.

Several processes have been studied for preparing thin films of high temperature materials. In general, these thin films include high amounts of lattice defects and also show poor adherence to the substrate due to their hardness.

5.3.1 SiC Thin Films

Various processes for making SiC films are available including vapor-phase reaction (214), plasma reaction (215), evaporation (216), rf-sputtering (217), and ion plating (218). Among these processes, one of the most convenient processes is rf-sputtering from a SiC target.

The crystalline structure of rf-sputtered SiC films varies from the amorphous to crystalline phase depending mainly upon the substrate temperature during deposition. Typical sputtering conditions are shown in Table 5.41.

Target	SiC ceramics (80 mm in diameter)
Sputtering gas	Argon (purity 99.9999%, 5 Pa)
Substrates	Fused quartz, silicon, alumina
Substrate temp.	200–500 °C
Target rf power (13.56 MHz)	1–3 W/cm²
Target–substrate distance	30 mm
Deposition rate	0.1–1 μm/h
Film thickness	4–5 μm

Table 5.41: Typical sputtering conditions for SiC thin films.

Figure 5.121 shows a typical surface structure and electron diffraction patterns of sputtered films on Si (111) substrates.

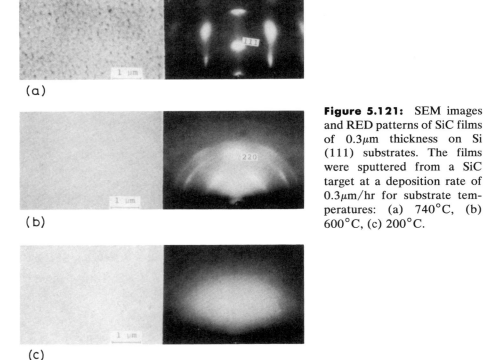

(a)

(b)

(c)

Figure 5.121: SEM images and RED patterns of SiC films of 0.3μm thickness on Si (111) substrates. The films were sputtered from a SiC target at a deposition rate of 0.3μm/hr for substrate temperatures: (a) 740°C, (b) 600°C, (c) 200°C.

Epitaxial β-SiC films on Si (111) substrates were obtained at a substrate temperature of 740°C as shown in Fig. 5.121(a). The epitaxial relationship is SiC(111) ∥ Si(111). Polycrystalline SiC films with the (220) plane parallel to the substrate surface were obtained for substrate temperatures higher than 550°C. A typical result is shown in Fig.

5.121(b). Amorphous SiC films with a specular surface were obtained below 500°C as shown in Fig. 5.121(c).

The crystalline films sometimes show the form with a hexagonal structure. A mixture of the α and β phases is also observed in these sputtered films. These sputtered films exhibit an infrared absorption band with a maximum at about 800 cm^{-1} which corresponds to the lattice vibration of bulk SiC (219), and also indicate the same mechanical hardness as the value of crystalline SiC.

The microhardness of sputtered SiC films is measured by pressing a diamond pyramidal indentor, such as used in the Vickers test, and measuring the diagonals of the square indentation. A typical scanning electron micrograph of the indentation for amorphous SiC films sputtered onto sapphire substrates is shown in Fig. 5.122 compared with those taken from the (001) sapphire substrate and (001) surface of a SiC single crystal. It shows that the diagonal of the indentation for the SiC film and therefore the hardness is smaller than that for the sapphire substrate, and nearly equal to that for the single crystal. Similar results were also observed with polycrystalline α-SiC films. Figure 5.123 shows the Vickers hardness calculated from the diagonal of the indentation for SiC films sputtered onto the substrate as a function of the indentor load. Surface hardness decreases with an increase in indentor load. At heavy loads of more than 100 g, the hardness becomes equal to that of sapphire substrates (1900 kgmm^{-2}) since the diamond indentor completely penetrates the SiC film on the sapphire. With a light load of less than 25 g, the surface hardness tends to increase to 4000kgmm^{-2}, corresponding to the hardness of the SiC layer. This value is nearly equal to that of bulk SiC.

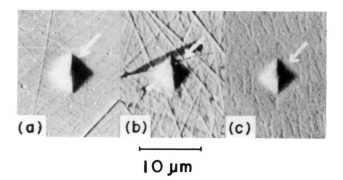

Figure 5.122: Typical scanning electron micrograph showing the indentation made by a diamond pyramidal indentor at an indentor load of 50g: (a) on (001) sapphire; (b) on SiC films about 2.6 μm thick sputtered onto a (001) sapphire substrate at 370°C with a deposition rate of 0.7μm/hr; (c) on an (001) SiC single crystal.

The wear resistance of sputtered SiC films is evaluated by a cyclical wear test. Table 5.42 shows typical results of the wear test for sputtered amorphous SiC films compared with the wear of pyrex glass and alumina plates (purity 97%). The wear of SiC films is much smaller than that of the pyrex glass and alumina plates. Similar results were also found for polycrystalline SiC films. The table suggests that sputtered SiC films are useful for hard surface coatings.

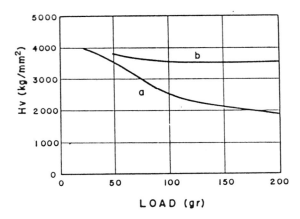

Figure 5.123: Vickers hardness as a function of indentor load: (a) for a SiC film about 2.6μm thick sputtered onto (001) sapphire at 370°C with deposition rate of 0.7μm/hr, (b) for (001) SiC single crystal.

Materials	Width of wear scar* (μm)	
	$v_s = 1$ mm sec^{-1}	$v_s = 4$ mm sec^{-1}
Pyrex glass	5.5	9
Alumina plate (purity 97%)	3.5	7
SiC film**	1	1

* Sliding distance 100 cm; load 4.0 g.
** Sputtered onto pyrex glass at 370 °C with a deposition speed of 0.7 μm h^{-1}.

Table 5.42: Cyclical wear test of a sputtered SiC film about 4.7μm thick for various sliding speeds v_s with a 0.7 mil. diamond stylus, compared with pyrex glass and an alumina plate.

Similar to SiC films, various kinds of rf-sputtered carbide films such as B$_4$C can be used for making hard surface coatings. However, as the hardness of the coating film increases, so does the internal stress contained in the film. This reduces the adherence of the film to the substrate, and an adhesion layer is necessary to make usable surface coatings. Figure 5.124 shows a cross section of a hard coating composed of a multilayer SiC/Si-C-O on a sapphire, Al$_2$O$_3$ substrate in which the Si-C-O layer acts as the adhesion layer. This multilayer is made by sputtering from a SiC target: The SiC target is first sputtered in a mixed gas of Ar and O$_2$ which results in the deposition of the adhesion layer, Si-C-O, and then the SiC hard coating is successively deposited by sputtering in pure Ar. The thickness distribution of Si, C, Al, and O atoms in the multilayer detected by the XMA is shown Figs. 5.125 and 5.126. A mutual diffusion layer exists between the sapphire substrate and the adhesion layer. Table 5.43 lists the composition and mechanical properties of hard coatings made by the rf sputtering process. The SiC-glass systems are prepared by rf sputtering from a pressed target of mixed SiC and borosilicate glass powder. The rf sputtered B$_4$C films show very poor adherence. To obtain surface coatings with high microhardness, mixed layer systems of SiC − B$_4$C are much more useful.

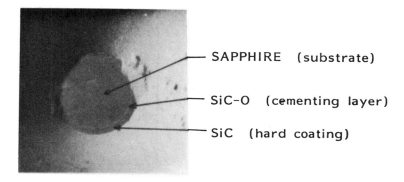

Figure 5.124: A cross section of a hard coating SiC layer on sapphire having cementing layer Si-C-O.

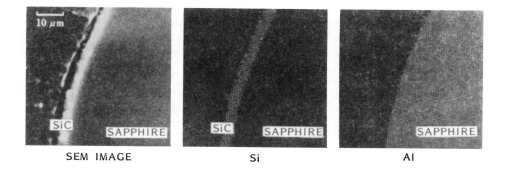

SEM IMAGE   Si   Al

Figure 5.125: Typical spatial distribution of Si and Al atoms detected by the XMA for the SiC hard coating.

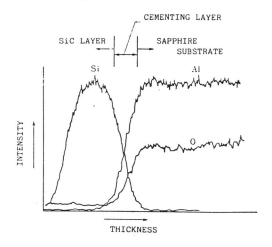

Figure 5.126: Typical thickness distribution of Si, O and Al atoms detected by the XMA for the SiC hard coating.

Composition	Sputtering target*	H_v (kg mm^{-2})	Wear resistance**	Remarks
SiC	Pressed SiC	4000	1	
SiC+1% glass		3800	0.9	Good adherence
SiC+5% glass	Mixture of SiC+ borosilicate glass	3300	0.1	
SiC+10% glass		2400	0.03	
SiC+25% B₄C	Mixture of SiC+B₄C	4500		
B₄C	Pressed B₄C	4800		Poor adherence

* Sputtering in argon at 300°–900 °C.
** Ratio of time required to a given wear volume against an iron plate (S-15C).

Table 5.43: Summary of properties if rf-sputtered hard coating films.

The electrical resistivity of sputtered SiC films is typically 2000Ωcm at room temperature. The temperature variations of film resistance are reversible when the substrate temperature during deposition is higher than the test-temperature range. Figure 5.127 shows typical temperature variations of the resistance of SiC films deposited on alumina substrate at a substrate temperature of 500°C measured at a temperature range between -100 to 450°C. It shows that the slope in the ln R vs. 1/T plot varies with temperature. The slope increases with the increase of temperature and the value lies between 1600 and 3400 K.

Sputtered silicon carbide thin films can be considered applicable for manufacturing silicon carbide thermistors as a high-temperature sensor instead of silicon carbide single crystals (220).

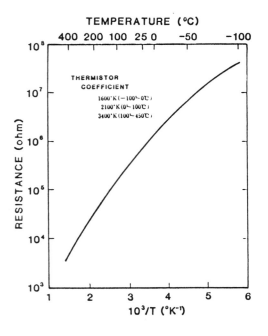

Figure 5.127: Typical temperature variation of the resistivity of sputtered SiC films about 2μm thick deposited onto alumina substrate; zero power resistance versus reciprocal absolute temperature.

The construction of the silicon carbide thin-film thermistor is shown in Figure 5.128. The thermistor is composed of silicon carbide thin film layers of 2 − 5μm thick, overlayed on substrate plates. The substrates are made of alumina ceramic of 0.6mm thickness. One pair of comb electrodes are inserted between the silicon carbide layers and the substrate. Fired Pt layers about 10μm thick are used as the comb electrode. The size of the substrate is 1x 8 x 0.6mm. The length and interval of the comb electrodes are 5 and 0.5mm, respectively.

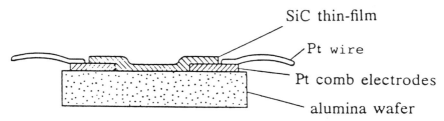

SiC thin-film

Pt wire

Pt comb electrodes

alumina wafer

Figure 5.128: Construction of SiC thin film thermistor; the thermistor is composed of sputtered SiC thin films on alumina substrate with fired Pt comb electrodes.

The thermistor is prepared by the following process: First, Pt comb electrodes are fired on an alumina wafer 70mm square. Then SiC thin films are deposited on the alumina wafer by rf-sputtering. The temperature of the alumina wafer is kept at 500- 550°C during sputtering. The typical deposition rate is 0.5 μm/h. Finally, the alumina wafer is annealed in air at about 550°C for 20 - 100 hours and then broken into thermistor tips. Figure 5.129 shows the alumina wafer and the thermistor tip. The Pt lead wires (0.3 mm in diameter) are welded to the fired Pt electrodes of the thermistor tip. The thermistor tip is, if necessary, packed in an envelope.

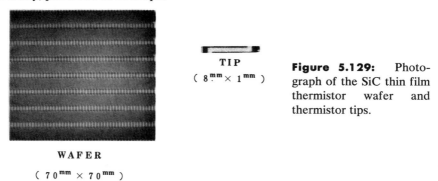

TIP

(8.mm × 1 mm)

Figure 5.129: Photograph of the SiC thin film thermistor wafer and thermistor tips.

WAFER

(70mm × 70mm)

SiC thin-film thermistors can be used for temperature sensing, temperature control, and flame detection with high reliability. They can operate between -100 and 450°C. The accuracy of temperature sensing or temperature control systems using these thermistors depends on the thermistor properties, i.e., thermistor resistance and thermistor coefficient. The accuracy of these properties are found to be 5% and 3%, respectively when alumina substrates and fired Pt electrodes are used. When one uses a silicon substrates and Cr/Au thin-film electrodes made by a photolithographic process, the accuracy of the thermistor resistance and thermistor coefficient are 1.5% and 0.5%, respectively. Figure 5.130 shows a typical SiC thin film thermistor for high-precision use made by a photolithographic process. The tip dimension is 0.5 x 0.5 x 0.1 mm and the time response is found to be less than 0.1s.

TIP

(0.9 mm x 0.9 mm)

WAFER (30 mm dia.)

Figure 5.130: SiC thin film thermistor tips formed on silicon wafer.

Typical thermistor properties of SiC thin film thermistors are listed in Table 5.44 together with those of SiC single crystals. The SiC thin film thermistor coefficient accuracy is much higher than that of the SiC single crystal (221).

	SiC thin film	SiC single crystal
Operating temperature range	−100–450°C	−100–450°C
Zero-power resistance	10 kΩ–1 MΩ at 25°C	2600 Ω
Zero-power accuracy	<±1.5% (thin-film electrode) <±5% (fired electrode)	±2%
Thermistor coefficient* B	1600 K (−100–0°C) 2100 K (0–100°C) 3400 K (100–450°C)	2000 K (25–125°C)
Accuracy	<±0.5% (thin-film electrode) <±3% (fired electrode)	±2.9%
Electrical stability	Resistance change <3% (400°C, 2000 h)	

* Average value.

Table 5.44: Typical characteristics of SiC thin film thermistors.

5.3.2 Tungsten Carbide (WC) Thin Films

Thin films of tungsten carbide have wide technological applications as wear resistant and protective coatings on a variety of surfaces such as cemented carbide tools, steel, copper, and copper alloys. For normal steel, copper, and copper alloys, a coating of WC-Co has been found to be suitable. The presence of cobalt is essential in WC coatings to reduce both friction and wear. However, tungsten carbide films with fine grains are highly adherent to steel substrate and do not necessitate any cobalt addition (40).

The thin films are prepared by direct sputtering from a tungsten carbide target, or by reactive sputtering of a tungsten target in a mixed gas of Ar and C_2H_2 as shown in Fig. 5.131. Typical sputtering conditions are shown in Table 5.45 (40).

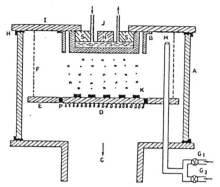

Figure 5.131: Schematic diagram of reactive r.f. magnetron system showing metallic belljar (A), target shield (B), vacuum pump (C), substrate heater (D), grounded base plane (E), perforated aluminum mesh (F), aluminum metallic partition (G), L gasket for vacuum sealing (H), top plate (I), target (J), substrate (K), opening of reactive gas (L), opening of inert gas (M), insulator ring (P), and precision needle valves G_1 and G_2 (Srivastava (1986) (40)).

Figure 5.132 shows X-ray diffraction patterns for tungsten carbide films sputtered on stainless steel at various substrate temperatures. It shows that a mixture of WC (cubic B_1), W_2C (hexagonal), and W_3C (A-15) cubic phase is formed at lower substrate temperatures ($\approx 200°C$). A single phase of WC is grown at higher substrate temperatures ($400 - 500°C$). These sputtered films consist of a randomly shaped granular surface with a grain size $400 \approx 500 \text{Å}$. A fractured cross section of the sputtered films show a columnar structure consisting of fine columns with a width of 300Å.

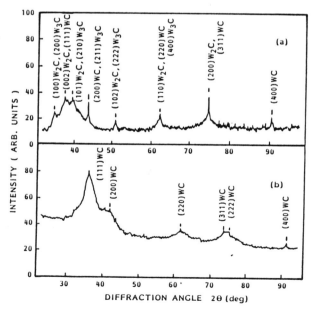

Figure 5.132: X-ray diffraction profiles of tungsten carbide films deposited on stainless steel at (a) 200°C and (b) 400°C (Srivastava (1986) (40)).

Under the normal conditions of reactive sputtering shown in Figure 5.131, carbides are known to form on the metal target surface and are subsequently sputtered off (normal mode, NM, deposition). Due to the low sputtering yield of these compounds, the rate of deposition of the corresponding film is low. The rate is the same order to that of direct sputtering of the compound target.

Figure 5.133 shows a modified geometry of the magnetron sputtering system (40). In the system the reaction of the sputtered species from the target with the reactive gas occurs only in the vicinity of the substrate surface, since separate zones of argon and acetylene are created by controlling the flow of the two gases such that the carbide formation on the tungsten target will be reduced. In this system tungsten carbide films have been deposited on the stainless steel substrate at rates as high as that of pure tungsten, $4.9\mu m/hr$, under sputtering conditions indicated in Table 5.45 (high rate mode, HRM of deposition).

Sputter system	planar magnetron
Target	tungsten
Sputter gas	Ar/C_2H_2
Gas pressure	2×10^{-2} Torr
Substrate	304 stainless steel
Sputter power	4.5 W/cm^2
Sub. temperature	$200 - 500^\circ C$
Growth rate	$0.36\mu m/hr$ [#]

[#] 4.9 μm/hr for the system shown in Fig.5.133.

Table 5.45: Sputtering conditions for the deposition of WC thin films.

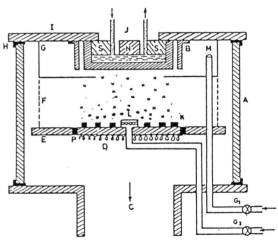

Figure 5.133: Schematic diagram of reactive r.f. magnetron system in high rate mode showing metallic belljar (A), target shield (B), vacuum pump (C), substrate heater (D), grounded base plate (E), perforated aluminum mesh (F), aluminum metallic metallic partition (G), L gasket for vacuum sealing (H), top plate (I), target (J), substrate (K), opening of reactive gas (L), opening of inert gas (M), insulator ring (P), and precision needle valves (G_1 and G_2) (Srivastava (1986) (40)).

Table 5.46 shows a summary of the composition and physical properties of these sputtered tungsten carbide films.

Mode of deposition	Substrate Temp. °C	Composition W	C	O	Crystallographic phase	Grain size (Å)	Micro-hardness kg/mm⁻²
NM	300	47	48	5	Single phase B1 f.c.c.	200	3200
NM	500	46	49	5	Single phase B1 f.c.c.	200	3200
HRM	300	58	37	5	Mixture of WC (hexagonal), $W_3C(A-15)$ and carbon (graphitic and diamond phase). Graphitic phase in excess.	300	2365
HRM	500	38	59	3	Mixture of WC (hexagonal), $W_3C(A-15)$ and and carbon (graphitic and diamond phase) Diamond phase in excess.	300	2365

Table 5.46: Composition, crystallographic structure and microhardness variation of WC thin films (Srivastava (1986) (40)).

5.3.3 Mo-Si Thin Films

Thin films of silicides, such as Mo-Si and Cr-Si, have high electrical resistivity and are useful for making thin film resistors. These silicide films are also used as Schottky gate materials for FET and interconnections for VLSI. Mo-Si films, for instance, are made by sputtering from a sintered Mo-Si target or a mixed powder of Mo and Si. Stable Mo-Si films are obtained at high substrate temperature, 500 to 600°C. Figure 5.134 shows the electrical properties for different compositions of Mo-Si. The figure suggests that the Mo-Si films show zero temperature coefficients of resistance for up to 80% Si film composition. These Mo-Si films are of technological interest as heating elements for a thermal head used in a thermal printer. A typical example of a thermal head is shown in Fig. 5.135.

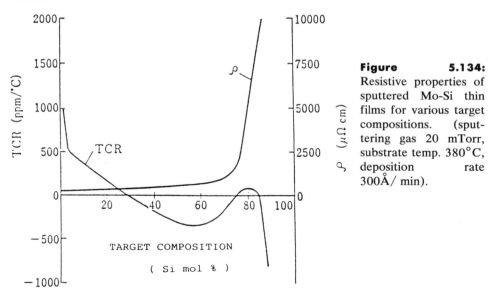

Figure 5.134: Resistive properties of sputtered Mo-Si thin films for various target compositions. (sputtering gas 20 mTorr, substrate temp. 380°C, deposition rate 300Å/ min).

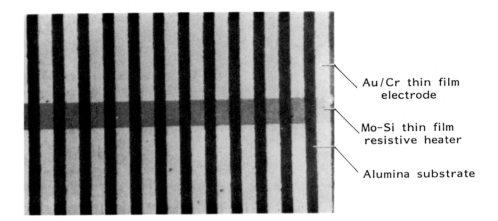

Figure 5.135: Mo-Si thin film thermal head for a printer: (8 dots/mm).

5.4 DIAMOND

Thin films of diamond are useful for making novel electronic devices. In the 1960's this kind of film was first deposited by decomposition of CH_4 in CVD system. In the 1970's several methods were considered for the deposition of diamond thin films, including plasma CVD, ion-beam deposition and sputter deposition. In the CVD process, thin films of diamond were prepared at the substrate temperature of 800 to 1000°C (222). Ion beam deposition and sputter deposition are attractive processes because it is possible to prepare thin films at room temperature due to their energetic adatoms.

Aisenberg and Chabot first tried to deposit thin films of diamond at room temperature by deposition of energetic carbon ions using ion beam deposition. The carbon ions were accelerated at 40 eV by a biased field (223). The resultant films were transparent with high electrical resistivity. Since their physical properties resemble diamond this kind of film was called diamondlike carbon (DLC) films. They may be composed of amorphous carbon with small diamond crystallites dispersed in the amorphous carbon network.

Sputter deposition of diamond films was first tried by Wasa and Hayakawa using rf-diode sputtering. They sputtered diamond powder in Ar and produced a transparent DLC film on a glass substrate at room temperature (224). The film showed poor adherence to the substrate due to their hardness as shown in Fig. 5.136.

In the 1980's, detailed studies were done on sputter deposition of DLC films. Weissmantel et al. deposited DLC films by sputtering from a graphite target using an ion beam sputtering system (225). These sputtered films can be used for optical hard coating in the infrared region.

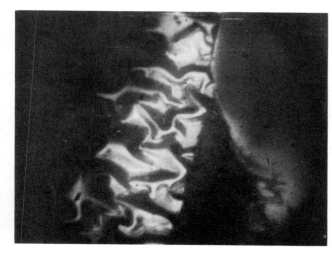

Figure 5.136: Micrograph of sputtered DLC thin film.

Bombardment of the substrate with hydrogen ions during ion beam sputtering deposition enhanced the growth of the diamond structure and reduced the graphite composition in DLC films (226).

The ion-beam sputter deposition system, which is used for the preparation of diamond films, is illustrated in Fig. 5.137. The graphite disk target (purity 99.999%, 100mm in diameter) was bonded to the water-cooled holder. An electron bombardment ion source was employed. The ion energy and ion current were 1200 eV and 60 mA, respectively. The ion-beam diameter was 25 mm. The incident angle of the ion-beam was about 30 degrees to the target. The substrate was placed near the target as illustrated in Fig. 5.137. The ion-beam sputtered the target and also bombarded the surface of the substrate at grazing incidence. The ion current densities were about 1mA/cm² and 0.04mA/cm² at the target and the substrate, respectively. The ion beam, which bombards the substrate, can modify the deposited carbon film. Table 5.47 summarizes the sputtering conditions.

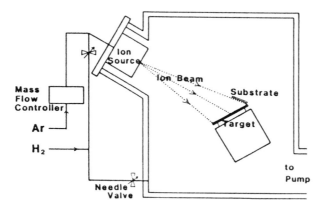

Figure 5.137: Construction of the ion-beam sputter deposition system for the deposition of diamonds.

Sputtering system	ion beam sputter
Target	graphite plate
	(100 mm in dia.)
Acceleration voltage	1200 V
Ion beam current	60 mA
Beam aperture	25 mm in diameter
Gas pressure	$5 \times 10^{-5} - 2 \times 10^{-4}$ Torr
	Ar/H_2 mixed gas
Substrate	(111)Si
Sub. temperature	RT -200°C
Target-ion source spacing	250 mm
Growth rate	0.3 - 0.4 µm/hr

Table 5.47: Sputtering conditions for depositing diamond thin films.

The optical transparency of the resultant films increased under irradiation of hydrogen ions, as shown in Fig. 5.138, and also increased their electrical resistivity. These results suggest that the graphite composition in the DLC films is reduced by bombardment of hydrogen ions. The structural analysis of DLC films are studied in detail by Raman scattering spectra.

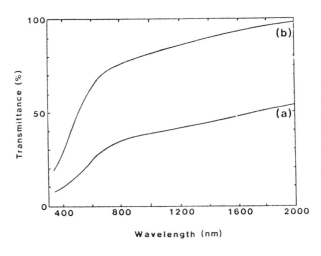

Figure 5.138: Optical absorption spectra of sputtered carbon films: (a) without hydrogen ion bombardment, (b) with hydrogen ion bombardment during deposition.

The most interesting phenomenon is that under bombardment of hydrogen ions several diamond crystals are partially grown on DLC films as shown in Fig. 5.139. The crystals exhibit the well defined morphology of cubic diamond and their lattice constant coincides with that of natural diamond.

Figure 5.139: Diamond crystals grown on the DLC thin films at room temperature.

The effects of bombarding the substrate with hydrogen and argon ions are considered as follows: Energetic activation and rapid quenching occur at the surface of the film during deposition. Bombardment with argon ions supplies the thermal and displacement spikes to the carbon atoms on the substrate. With increased pressure these spikes may create high-temperature regions. The microcrystals of diamond are partly formed around these spikes. These diamond regions barely grow in the film, because they are surrounded by nondiamond regions. The carbon atoms in these nondiamond regions are weakly bonded and are easily activated by hydrogen ion bombardment and change to the gas of C_xH_y. The nondiamond regions may then be selectively removed.

The diamond regions, which are formed by these mechanisms, have many broken bonds which make the structures highly nonequilibrium and weak. If the broken bonds are rapidly compensated by hydrogen, the diamond regions are quenched and may grow to a stable size. The bombarding hydrogen ions are so active that compensate the broken bond and activate the growth of diamond.

5.5 SELENIDES

Thin films of selenides such as ZnSe are prepared by rf-sputtering from ZnSe cathode in an Ar atmosphere. Typical sputtering conditions are shown in Table 5.48 (3). In the rf-sputtering system the anode is perforated and substrates are placed behind the anode at floating potential so as to reduce the effects of the high energy sputtered atoms and secondary electrons from the cathode.

Sputter system	Rf-diode (13.56MHz)
Target	ZnSe disk (5N)
Sputter gas	5×10^{-3} Torr (Ar 5N)
Spacing, target-anode	20mm
target-substrate	27mm
Rf-power	$0.1 \sim 2mA/cm^2$ $-900 \sim -1500V$
Substrate	(100)NaCl , (111),(100)GaAs, (111)Si
Growth rate	$0.03 \sim 1$ m/hr

Table 5.48: Typical sputtering conditions for depositing ZnSe thin films.

Transmission electron diffraction patterns from ZnSe layers sputtered onto the (100) plane of NaCl single crystals exhibit the polycrystalline or epitaxially grown monocrystalline phase as shown in Fig. 5.140. The monocrystalline films consist of a single phase of cubic ZnSe structure with the (100) plane parallel to the NaCl surface, but the polycrystalline films consist of two phases, a cubic ZnSe structure and a hexagonal ZnSe structure. The degrees of structural ordering depend on deposition rates and substrate temperatures as shown in Fig. 5.141. In this figure, a minimum epitaxial temperature T_e was found to exist for a given deposition rate R, and the exponential relationship $R = A \exp(-Q_d/kT_e)$ was obeyed for epitaxy. Here, A is a constant and Q_d is the energy of surface diffusion of adatoms. In the present case we have $Q_d \approx 0.23eV$. The satellite spots in Fig. 5.137 denote the existence of twins in the epitaxial layers. The crystalline size of the sputtered layer was in the order of 0.03 to 0.3 μm as seen in the microstructure.

It is noteworthy, however, that single crystal films are epitaxially grown at a low substrate temperature of 150°C, although bulk ZnSe single crystals are synthesized at about 1,000°C. This suggests that the sputtering process lowers the synthesizing temperature of crystals. Thin films of ZnSe are promising materials for making heterojunction photodiodes, solar cells, and blue light emitting diodes.

Figure 5.142 shows a typical oscilloscope trace of a current-voltage characteristic for the ZnSe/Si diode measured in darkness at room temperature. The diode shows rectifying properties and the direction of rectification suggests that the ZnSe layer is a n-type semiconductor. As shown in Figure 5.143, the forward current-voltage characteristics of the diodes exhibit ohmic currents at a low applied voltage and space charge limited currents (SCLC) at a high applied voltage. The SCLC exhibited trap-controlled behavior similar to that occurring in vacuum deposited n-type ZnSe - p-type Ge heterojunction diodes in which the SCLC flowed in a high resistivity ZnSe layer (227). The high resistivity may result from defects produced by the bombardment of energetic particles and/or inclusions of impurity gases during deposition.

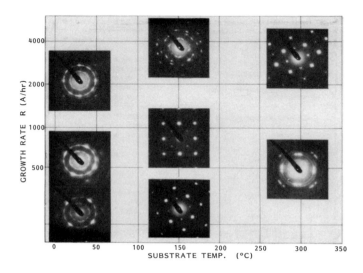

Figure 5.140: Transmission electron diffraction patterns for ZnSe films sputtered on (100) NaCl substrates.

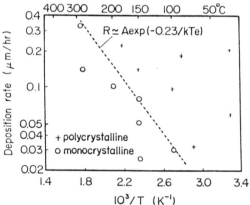

Figure 5.141: Deposition rate vs. reciprocal substrate temperature showing the transition from the polycrystalline to monocrystalline structures for ZnSe films sputtered onto (100) NaCl substrates.

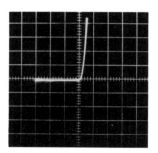

Figure 5.142: Dark current-voltage characteristic for sputtered n-type ZnSe p-Si heterojunction diode: vertical scale: $10\mu A/div$. Horizontal scale: $1V/div$.

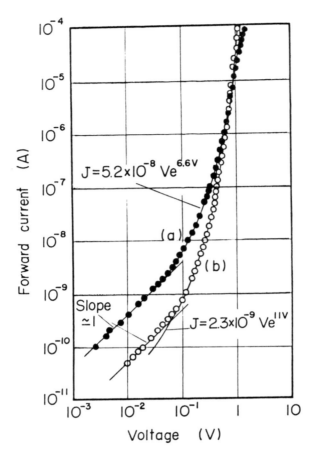

Figure 5.143: Forward current-voltage characteristics for two sputtered diodes measured at room temperature in a dark space: (a) n-type ZnSe p-type Si, (b) n-type ZnSe p-type GaAs diodes.

Several processes have been studied for making ZnSe single crystal films with low defect density. The MBE and MOCVD processes are considered as the most promising processes (230,231). It is also reasonable to consider that a lower defect density may be achieved by controlling the impinging particle energy during sputter deposition.

5.6 AMORPHOUS THIN FILMS

Amorphous materials are prepared by quenching of melts or vapors (228-231). Table 5.49 summarizes these quenching processes. Among these processes, the vapor quenching is achieved by many thin film deposition processes including vacuum deposition, sputter deposition, and CVD. Thin films of the amorphous phase are deposited at a substrate temperature below the crystallization temperature of the thin films. Table 5.50 shows the crystallization temperature for various materials. The metals generally show low crystallization temperature, however compounds such as oxides typically are much higher (232,233).

Quenching process		Quenching speed [deg./sec.]
from melt	annealing	$10^{-2} \sim 10^{-5}$
	air-quenching	10
	liquid-quenching	10^3
	splutter-quenching	10^5
	roller-quenching	$10^6 \sim 10^8$
from vapour	vacuum deposition , CVD	$>10^9$
	sputtering	$>10^{16}$

Table 5.49: Summaries of quenching processes.

Materials	Crystallization temp. (K)	Materials	Crystallization temp. (K)
V	3~4	Cr_2O_3	718
Cr	~220	MgO	598
Ga	10~60	NiO	558
Ge	743	Al_2O_3	1003
Si	993	Fe_2O_3	808
Bi	10~30	GaAs	~603
Se	~300	SiO_2	948
Te	~280	TiO_2	753
Sn-Cu	~60	Ta_2O_5	1013

Table 5.50: Crystallization temperature.

The crystallization temperature of metals will increase due to the inclusion of residual gas during deposition. For instance, the crystallization temperature of Fe thin films deposited at $10^{-12} - 10^{-10}$ Torr is 4 K. The temperature increases to 75 K when the O_2 partial pressure is 10^{-8} Torr. A typical sputtering system for the deposition of amorphous thin films is shown in Figure 5.144.

The temperature rise of the substrates during deposition is 200 to 500°C in the conventional diode sputtering system when the substrates are not cooled by water. The substrates are cooled by water, liquid nitrogen, or He for making amorphous thin films. Figure 5.145 shows a photograph of the sputtering system for the deposition of amorphous thin films. The magnetron target is useful in reducing the temperature rise of the substrates.

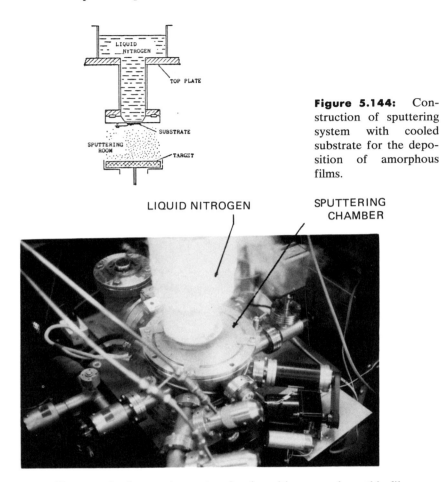

Figure 5.144: Construction of sputtering system with cooled substrate for the deposition of amorphous films.

Figure 5.145: Photograph of sputtering system for depositing amorphous thin films.

In sputter deposition the energy of adatoms is of the order 1 to 10 eV. Thus the quenching time of these adatoms is estimated to be higher than $10^{16}K/$ sec since they will lose their energy in 10^{-12} sec, which corresponds to the frequency of the thermal lattice vibration of the substrate surface atoms. The quenching rate expected in sputter deposition is much higher than the value obtained in quenching from melts as indicated in Table 5.49. This suggests that exotic structures may be prepared by sputter deposition process.

5.6.1 Amorphous ABO₃

There has been a growing interest in amorphous states of ABO_3 type ferroelectric materials for both their physics and practical applications. The possibility of ferroelectricity in amorphous dielectrics was shown theoretically by Lines (234). Experimental confirmation of this possibility is being explored. Glass et al. prepared amorphous LiNbO$_3$ and LiTaO$_3$ by a roller-quenching method and reported dielectric anomalies with a peak of $\varepsilon_r > 10^5$ near the crystallization temperature (235). These amorphous materials exhibit other interesting properties such as pyroelectricity and high ionic conductivity (236). Amorphous PbTiO$_3$ was prepared by Takashige et al. using a modified roller-quenching method and their dielectric anomalies were reported (237).

Sputter deposition on cooled substrates is an interesting method for the preparation of amorphous ABO$_3$ materials. Thin and uniform layers with a large area are easily obtained. The cooling of substrates during the deposition suppresses crystallization of the deposited materials and makes it possible to produce amorphous materials.

Thin films of amorphous LiNbO$_3$ are made by rf-sputtering from a LiNbO$_3$ compound target (238). Typical sputtering conditions are shown in Table 5.51.

Sputtering system	rf-diode
Target	LiNbO$_3$ powder
Sputter gas	Ar/O$_2$ (1:1)
Gas pressure	5 X 10^{-3} Torr
Sub.temperature	water-cooling
Substrate	glass
Rf-power	2.5 W/cm^2
Dep.rate	0.38 μm/hr
Film thickness	0.5 - 1.0 μm

Table 5.51: Sputtering conditions for the preparation of amorphous LiNbO$_3$ thin films.

An rf-planar magnetron sputtering apparatus is employed for preparation of the films. A stainless steel dish 10cm in diameter and filled with LiNbO$_3$ powder is used as a target. The LiNbO$_3$ powder is synthesized from Li$_2$CO$_3$ (purity 99%) and Nb$_2$O$_5$ (purity 99.9%). Substrates are polished fused quartz. Before the deposition of LiNbO$_3$, Au/Cr stripes are vacuum evaporated on the substrates as bottom electrodes for the measurement of dielectric properties. The substrate is attached to a water-cooled holder so as to avoid crystallization of the LiNbO$_3$ film during the sputter deposition.

Transparent and colorless films with smooth surfaces were obtained by sputter deposition. Figure 5.146(a) shows an X-ray diffraction pattern of the sputtered film about 1 μm thick without electrodes. The pattern exhibits halo-diffraction, which indicates that the film is in an amorphous state. In order to examine the crystallizing temperature, the films were annealed in air. From the X-ray diffraction pattern, it was found that the film annealed at 350°C for 1 hr was still amorphous. Further annealing at 500°C for 1 hr caused crystallization of the film. Figure 5.146 (b) shows an X-ray diffraction pattern of the annealed film, which agrees well with that of LiNbO$_3$ powder (ASTM card 20-631) shown in Figure 5.146(c). The film annealed at 500°C is considered to be polycrystalline LiNbO$_3$.

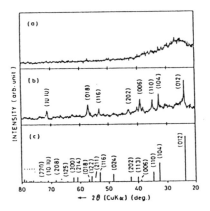

Figure 5.146: X-ray diffraction patterns of (a) as-grown film, (b) film annealed in air at 500°C for 1 hr, and (c)LiNbO₃ powder (ASTM card # 20-631).

Figure 5.147 shows a typical temperature dependence of dielectric properties for as-sputtered amorphous LiNbO₃ films. In this figure, the ε_r of single crystal LiNbO₃ along the a-axis and c-axis at 100 kHz reported by Nassau et al. are plotted in dotted lines (239). These dielectric properties differ from those of the roller-quenched samples: the dielectric anomalies are observed at about 350°C, which is 200 ≃300°C lower than that of the roller-quenched LiNbO₃.

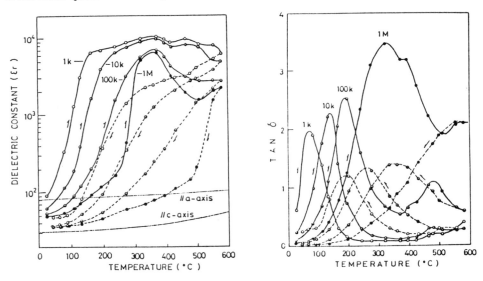

Figure 5.147: Temperature dependence of dielectric properties for the first cycle of heating and cooling of the as-grown LiNbO₃ films in amorphous state measured at 1k, 10k, and 1 MHZ. Dotted lines indicate ε_r of single-crystal LiNbO₃ along a-axis and c-axis at 100 KHZ.

Thin films of amorphous $PbTiO_3$ are also prepared by rf-sputtering from a sintered $PbTiO_3$ target on cooled substrates (240). In this case there are Pb metal crystallites in as-sputtered $PbTiO_3$ films, and these films exhibit a high surface electrical conductivity of $\simeq 10(\Omega cm)^{-1}$ due to hopping of electrons between the Pb metal crystallites. When the as-sputtered films are annealed above 220°C, the surface electrical conductivity is reduced to $\leq 10^{-4}(\Omega cm)^{-1}$. The annealing causes a decrease in Pb metal crystallites in the film. The dielectric constant of the annealed $PbTiO_3$ shows broad maxima at 440°C. When the films are annealed above 500°C, they become polycrystalline $PbTiO_3$ films.

These maxima, sometimes observed in the roller-quenched $PbTiO_3$ (241), are similar to the temperature dependence of the dielectric constants of PbO (red) and are attributed to the presence of PbO (red) crystallites. Broad maxima are observed at 400°C in cooling. These peaks may be blurred dielectric anomalies which are attributed to nonstabilized ferroelectric ordering. This was also observed in the roller-quenched $PbTiO_3$(237). These results suggest that the structure of the annealed film resembles that of roller-quenched amorphous $PbTiO_3$.

5.6.2 Amorphous SiC

The crystallization temperature of SiC thin films sputtered from SiC target is about 500°C, below which the films are amorphous. However, the stable amorphous phase is obtained in SiC thin films reactively sputtered from a Si target in a C_2H_2 or CH_4 gas (242). Typical sputtering conditions for the deposition of amorphous SiC films are shown in Table 5.52. Although the substrate temperature is higher than the crystallization temperature of conventional rf-sputtered SiC thin films, the reactively sputtered SiC thin films show the amorphous phase.Their infrared transmission spectra are shown in Fig. 5.148. The large absorption band at $800 cm^{-1}$ is due to the fundamental lattice vibration of SiC (243). The small absorptions due to Si-H and C-H are superposed on the spectra. The inclusion of hydrogen atoms in sputtered SiC films stabilize the amorphous phase.

Sputtering system	rf-diode
Target	Si 6N 100 mm in dia.
Sputter gas	Ar / CH_4, C_2H_2
Gas pressure	2×10^{-3} Torr
Substrate	Si (111)
Sub.temperature	200–740°C
Rf-power	3 –4W/cm^2
Dep. rate	0.2 – 0.6 μm/hr
Film thickness	0.8 – 2 μm

Table 5.52: Sputtering conditions for the preparation of amorphous SiC thin films.

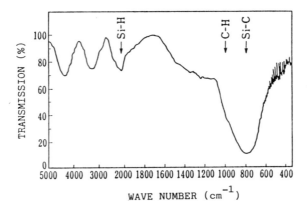

Figure 5.148: Infrared transmission spectra of sputtered amorphous SiC thin films.

Thin films of amorphous Si are also deposited by the sputtering process (244).

5.7 SUPER-LATTICE STRUCTURES

Intercalated structures consisting of thin alternating layers grown by vacuum deposition such as the MBE process (245), chemical vapor deposition including the ARE process (246), and sputtering deposition (247). Among these processes the sputtering process allows the preparation of thin alternating layers of refractory metals and compounds such as oxides, nitrides, and III-V alloys. As described previously alternating layers of compound oxides including perovskite dielectrics and superconductors are made by sputtering from multi-targets. Greene et al. deposited the III-V compound alternating layers of InSb/GaSb/InSb..... (248). In multitarget sputtering the substrate is continuously rotated through two or more electrically and physically isolated sputtering discharges. The structural properties depend on the layer thickness deposited per target pass and the rate of interlayer diffusion. Eltoukhy et al. have deposited the InSb/GaSb superlattice structures with layer thicknesses ranging from 12.5 to 50Å at a substrate temperature of 200 to 250°C (249). They have suggested that the superlattice structures only a few monolayers thick may easily be grown by sputtering from the multitarget even at relatively high substrate temperatures.

Superlattice structures are also deposited by reactive sputtering. Figure 5.149 shows the construction of the sputtering system for making thin alternating layers of Nb/NbN. For the first layer, the Nb target is sputtered in Ar, later, a mixed gas of Ar/N_2. Periodic change of the sputtering gas composition results in the formation of the superlattice Nb/NbN. Typical sputtering conditions are shown in Table 5.53. The Auger depth profile of the layered structure is shown in Fig. 5.150. The thickness of each layer is about 100Å. The period of the layered structure corresponds to the variations of the sputtering gas composition. This sputtering process achieves a layer thickness of thinner than 10Å. Contamination by impurities can be reduced when the ion beam sputtering system is used for deposition. Typical structural properties are shown in Figure 5.151 (250).

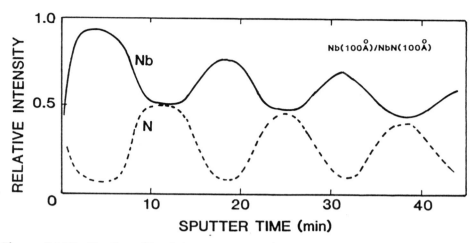

Figure 5.149: Depth profile of the sputtered Nb/NbN superlattice structure measured by AES.

	Nb thin layers	NbN thin layers
Sputtering gas	Ar(3X10^{-2}Torr)	Ar(3X10^{-2}Torr) N$_2$(0.75X10^{-2}Torr)
Substrate temp.	300°C	300°C
Deposition rate	50,100 A/min.	100 A/min.

Table 5.53: Typical sputtering conditions for the deposition of the superlattice structure Nb/NbN.

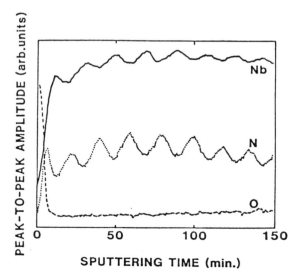

Figure 5.150: Depth profile of the sputtered Nb/NbN superlattice structure measured by AES.

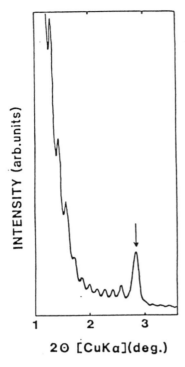

Figure 5.151: The X-ray diffraction pattern of the sputtered Nb/NbN superlattice structure.

The sputtering process is also a promising method for making the superlattice structure for refractory metals, oxides, and nitrides which are useful for magnetic media, superconductors, dielectrics, and X-ray mirrors. However, energetic adatoms will produce lattice defects for which one will encounter some difficulty in making semiconducting devices.

5.8 ORGANIC THIN FILMS

Sputtered particles from an organic polymer target will be composed of atomic species and/or monomers of the target materials. Thin films of the organic polymer may be deposited since these sputtered particles may be polymerized in the sputtering gas discharge or on the substrate during film formation.

The organic polymer film is used as the target which is fixed onto a cooled copper target holder with silicon grease so as to reduce the temperature rise of the target surface. Figures 5.152 and 5.153 show the infrared transmission spectra of organic thin films sputtered from a teflon and a polyimide target in Ar. The sputtered teflon films exhibit the C-F absorption band corresponding to the bulk teflon. The films are transparent and their physical properties resemble bulk teflon. However, the sputtered polyimide films show different absorption from bulk polyimide, and thus may be significantly far different from the bulk polyimide.

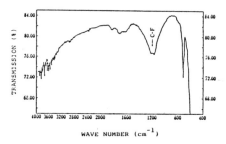

Figure 5.152: Infrared transmission spectra of sputtered teflon thin films.

5.9 MAGNETRON SPUTTERING UNDER A STRONG MAGNETIC FIELD

Under a strong magnetic field (1,000G) a high discharge voltage (1000V) is kept in the magnetron sputtering system. The sputtering system operates at a low gas pressure of 10^{-4} to 10^{-5} Torr. The high sputtering voltage and low working pressure will cause the impingement of high speed sputtered atoms on the substrates. This may result in the lowering of the growth temperature as described in the synthesis of $PbTiO_3$ thin films. Unusual properties are also observed in these sputtered films. This section presents some of the interesting phenomena observed in sputtered films prepared by magnetron sputtering under a strong magnetic field.

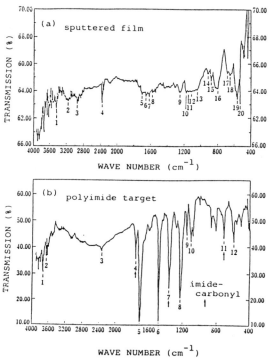

Figure 5.153: Infrared transmission spectra of sputtered polyimide thin films (a) and polyimide target (b).

5.9.1 Abnormal Crystal Growth

Polycrystalline ZnO films of hexagonal structure are prepared on a glass substrate by dc- or rf-sputtering from a zinc or ZnO target in an oxidizing atmosphere. Table 5.54 shows typical sputtering conditions and the crystallographic structure of ZnO films prepared in a conventional dc-sputtering system and in a dc-magnetron sputtering system. In the conventional sputtering system where the working pressure is 10 to 100 mTorr , it can be seen that the c-axis is preferentially oriented normal to the film surface, i.e. the (002) plane is parallel to the film surface. When ZnO films are prepared at a low working pressure of 1 mTorr or less in the magnetron sputtering system, the c-axis is predominantly parallel to the film surface, i.e. the (110) or (100) plane is parallel to the film surface (251). Typical electron micrographs and electron diffraction patterns of these sputtered films are shown in Fig. 5.154. It shows that microstructures of ZnO films prepared by magnetron sputtering exhibit a pyramidal pattern and clearly differ from those prepared by conventional sputtering. The microstructures of ZnO films prepared by conventional sputtering are composed of small hexagonal crystallites.

The c-axis orientation obtained in conventional sputtering is explained by the fact that surface mobility of adatoms is high during film growth, and sputtered films obey the empirical law of Bravais (252). The change in crystallographic orientation with these sputtering systems may be related to the difference in their working pressures. At low working pressure, the oxidation of the zinc cathode will not be completed during sputtering. Moreover, the low working pressure causes the impingement of high energy sputtered zinc atoms and/or negative oxygen ions on the substrates. This may lead to unusual nucleation and film growth processes in which Bravais' empirical law is inapplicable.

Sputtering system[a]	Sputtering pressure $(10^{-3}$ Torr)	Substrate temperature (°C)	Deposition rate $(\mu m\ h^{-1})$	Film thickness (μm)	Crystallographic orientation[d]
Conventional	35[b]	40	0.03	0.1	$C_{\perp}$
d.c. diode		100	0.15	0.3	$C_{\perp}$
		200	0.03	0.1	$C_{\perp}$
		200	0.15	0.3	$C_{\perp}$
		200	0.3	0.3	$C_{\perp}$
		300	0.03	0.1	$C_{\perp}$
		300	0.15	0.3	$C_{\perp}$
		300	0.3	0.3	$C_{\perp}$
D.c.	1[c]	40	0.03	0.1	$C_{\perp}$
magnetron		40	0.12	0.36	$C_{\perp}+C_{//}$
		70	0.7	0.35	$C_{//}$
		150	0.1	0.3	$C_{//}$
		150	0.7	0.3	$C_{//}$
		200	0.7	0.3	$C_{//}$
		270	0.07	0.2	$C_{//}$
		270	0.6	0.3	$C_{//}$

[a] Pure Zn cathode, 7059 glass substrates.
[b] 50% oxygen and 50% argon.
[c] 30% oxygen and 70% argon.
[d] $C_{\perp}$, c axis normal to the film surface; $C_{//}$, c axis in the film surface.

Table 5.54: Crystallographic orientation of polycrystalline ZnO thin films.

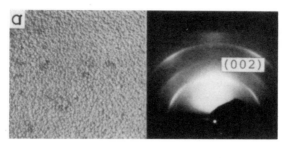

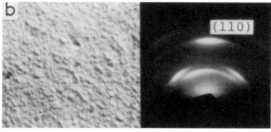

Figure 5.154: Electron micrographs and electron diffraction patterns for ZnO thin films 5000 Å thick on 7059 glass: (a) sputtered in a conventional diode system; (b) sputtered in a magnetron system.

5.9.2 Low Temperature Doping of Foreign Atoms Into Semiconducting Films

Co-sputtering of foreign atoms seems to be useful for controlling the electrical conductivity of semiconducting films during sputtering deposition. Table 5.55 shows typical experiments for polycrystalline ZnO thin films in various sputtering systems. In the experiments an aluminum or copper auxiliary cathode was co-sputtered with a zinc main cathode in an oxidizing atmosphere (251).

Sputtering system[a]	Sputtering pressure $(10^{-3}$ Torr)	Content of foreign metals (at.%)	Substrate temperature (°C)	Deposition rate $(\mu m\ h^{-1})$	Film thickness (μm)	Dark conductivity $(\Omega^{-1}\ cm^{-1})$
Conven-	60[b]	0	300	0.25	0.5	1.6×10^{-6}
tional		0.2 (Al)	300	0.075	0.15	5×10^{-6}
d.c. diode		0.2 (Cu)	300	0.1	0.2	1.9×10^{-6}
D.c.	1[c]	0	40	0.03	0.1	1×10^{-6}
magnetron		0	200	0.7	0.3	1×10^{-4}
		1.3 (Al)	200	1.2	0.6	8×10^{-2}
		0.5 (Cu)	200	0.9	0.45	3×10^{-8}

[a] Pure zinc cathode, 7059 glass substrates.
[b] 30% oxygen and 70% argon.
[c] 50% oxygen, 50% argon.

Table 5.55: Electrical conductivity of polycrystalline ZnO thin films.

It shows that in the conventional sputtering system, co-sputtering of aluminum or copper hardly affects the conductivity of the resultant films, while in the magnetron sputtering system the co-sputtering of aluminum or copper strongly affects the conductivity : aluminum increases the conductivity by over three orders of magnitude and copper decreases it by approximately the same factor. This suggests that in magnetron sputtering aluminum is probably introduced as a donor and copper as an acceptor or a deep trap. In conventional sputtering the co-sputtered atoms stay mainly at the crystal boundaries in the sputtered films and are probably not incorporated into the crystal lattice. Figure 5.155 shows the typical temperature variation of dark conductivity for ZnO films prepared by magnetron sputtering. The conductivity is controlled in the range $10^{-1} - 10^{-8}\Omega^{-1}cm^{-1}$ by doping with foreign atoms in the co-sputtering process.

Optical absorption measurements suggest that the forbidden gap width of these films is 3.3 eV, and that the acceptor or deep trap level of copper is 2.5 eV below the conduction band at room temperature. The temperature dependence of the carrier concentration suggests that the donor level of aluminum is 0.08 eV below the conduction band. The doping of foreign atoms by the co-sputtering process may possibly be the result of the impingement of high energy sputtered atoms during film growth.

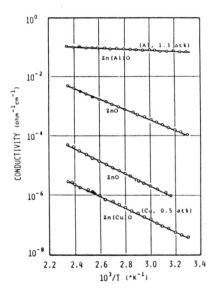

Figure 5.155: The temperature variation of the dark conductivity for ZnO films with and without co-sputtered foreign metals prepared in a magnetron sputtering system.

The highly conductive aluminum doped ZnO films can be utilized for making ZnO/Si heterojunction photodiodes and switching diodes (257). Figures 5.156 and 5.157 show a typical dark current-voltage characteristic and photoresponse of the photodiode respectively. The diode was prepared by depositing an n-type aluminum doped ZnO layer onto a p-type (111) silicon single crystal wafer at about 200°C. The resistivity and thickness of the ZnO layer were 10 to 100Ωcm and 0.6μm respectively. The resistivity of the silicon wafer was 20Ωcm. The diode shows a photovoltage of about 100mV for an open circuit and a photocurrent of 1mA/cm² under irradiation from a tungsten lamp (2800°C) of 10^4lx.

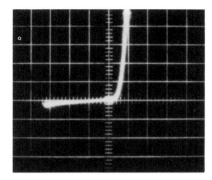

Figure 5.156: The dark current-voltage characteristic for a sputtered n-p ZnO(Al)O/Si photodiode: the vertical scale is 100μA per division and the horizontal scale is 2 volts per division.

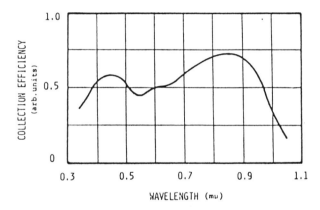

Figure 5.157: The spectral response of the collection efficiency for an n-p Zn(Al)O/Si photodiode.

It is now believed that the magnetron sputtering system can be useful in industry for metallization of semiconductor devices because of its high deposition rate. However, the system offers even more attractive features when thin films of compounds such as oxides, nitrides and carbides are deposited. Some experimental results obtained with the dc-magnetron sputtering system are summarized in Table 5.56 (253). Further studies will undoubtedly bring success in the formation of unusual materials which have yet to be obtained by vacuum deposition processes.

Composition of sputtered films	Sputtering conditions				
	Target	Sputter gas (Torr)	Substrate[b] temperature (°C)	Deposition rate R (μm h⁻¹)	Rate factor K[c] (μm h⁻¹ per V A cm⁻²)
Al_2O_3	Al	$Ar + O_2$ $p_{O_2} > 1 \times 10^{-4}$	≈ 200	0.5–1.0	0.24[d]
Ta_2O_5	Ta	O_2	≈ 200	0.1–0.3	0.11[e]
TiO_2	Ti	$Ar + O_2$ $p_{O_2} > 3 \times 10^{-5}$	≈ 200	0.1–0.3	0.15[f]
PbO	Pb	$Ar + O_2$ $p_{O_2} \approx (2–3) \times 10^{-4}$	40–200	2.0–6.0	3.0
PbO_2	Pb	O_2	≈ 40	2.0	2.0
ZnO	Zn	$Ar + O_2$ $p_{O_2} \approx (3–5) \times 10^{-4}$	40–270	0.03–1.0	0.3
ZrO_2	Zr	$Ar + O_2$ $p_{O_2} \approx 3 \times 10^{-4}$	≈ 200	0.1–0.3	0.3
Pb–Ti–O	Pb–Ti	$Ar + O_2$ $p_{O_2} \approx 3 \times 10^{-4}$	100–300	0.3	0.3
AlN	Al	$Ar + N_2$ $p_{N_2} \approx 3 \times 10^{-4}$	≈ 150	0.3	0.6
ZrN	Zr	$Ar + N_2$ $p_{N_2} \approx 3 \times 10^{-4}$	≈ 150	0.7–2	0.42
TiN	Ti	$Ar + N_2$ $p_{N_2} \approx 3 \times 10^{-4}$	≈ 150	0.6–1.8	0.3
Ti–Zr–N	Ti–Zr	$Ar + N_2$ $p_{N_2} \approx 3 \times 10^{-4}$	≈ 150	0.6–1.8	0.4
Ti–Al–N	Ti–Al	$Ar + N_2$ $p_{N_2} \approx 3 \times 10^{-4}$	≈ 150	0.6–1.8	0.4
Ti–Zr–Al–N	Ti–Zr–Al	$Ar + N_2$ $p_{N_2} \approx 3 \times 10^{-4}$	≈ 150	0.6–1.8	0.4

[a] Coaxial geometry; cathode current, 0.1–5 mA cm^{-2}; discharge voltage, 300–1000 V; cathode diameter, 20 mm; anode diameter, 60 mm; total sputtering gas pressure, $(0.6–1) \times 10^{-3}$ Torr; magnetic field, 1000–7000 G; substrates at anode.

[b] 7059 glass substrates.

[c] $K = R/V_s i_s$ where V_s is the discharge voltage and i_s is the cathode current density.

Table 5.56: Summary of properties of compound films prepared by dc magnetron sputtering in a strong magnetic field.

Film properties		Miscellaneous
Structure; thickness (μm)	Electrical properties	
Amorphous; 0.1-0.3	$\varepsilon^* \approx 8$	
Amorphous; 0.1-0.3	$\varepsilon^* \approx 20$-30 $\tan \delta \leqslant 0.01$ $V_b \approx 0.5$-1 MV cm^{-1}	
Polycrystal, rutile, anatase; 0.1-0.5	$\varepsilon^* \approx 20$-60 $\tan \delta \approx 0.01$-0.01 $V_b \approx 0.3$-0.7 MV cm^{-1} TCC ≈ -300 to $+300$ ppm °C^{-1}	Optical absorption edge, 3.4 eV; at high p_{O_2} strongly sensitive to water vapour
Polycrystal, red; 0.1-1	$\varepsilon^* \approx 20$-25 $\tan \delta \approx 0.04$-0.06 $V_b \approx 0.2$-0.5 MV cm^{-1} $\rho \approx 10^4$-$10^5 \Omega$ cm TCC $\approx +700$ to $+900$ ppm °C^{-1}	Optical absorption edge, 2.7 eV; photoconductive, max. at 2.7 eV; dielectric anomaly at 470 °C
Polycrystal; 0.1-1	$\rho \approx 10^{-2}$-$10^{-3} \Omega$ cm	At high substrate temperature the composition becomes PbO
Polycrystal, $c_{//}$ $(c_\perp)$; 0.1-1	$\varepsilon^* \approx 8$-10 $\rho \approx 10^4$-$10^6 \Omega$ cm	Optical absorption edge, 3.2 eV; photoconductive; piezoelectric
Amorphous; 0.1-0.3	$\varepsilon^* \approx 13$-20 $\tan \delta \approx 0.003$-0.007 $V_b \approx 0.5$-1 MV cm^{-1} TCC $\approx +100$ to $+300$ ppm °C^{-1}	
Mixture of PbO, TiO$_2$, PbTiO$_3$ with Pb:Ti ≈ 1; 0.1-1.0	$\varepsilon^* \approx 120$ $\tan \delta \approx 0.005$-0.015 $V_b \approx 1$-1.5 MV cm^{-1} TCC ≈ -100 to $+100$ ppm °C^{-1}	Dielectric anomaly at 490 °C
Amorphous; 0.1-0.5		
Polycrystal; 0.1-0.5	$\rho \approx 1000 \mu\Omega$ cm TCR ≈ -200 ppm °C^{-1}	
Polycrystal; 0.1-0.3	$\rho \approx 250 \mu\Omega$ cm TCR ≈ 150 ppm °C^{-1}	
Polycrystal, solid solution of Ti-Z-N with Ti:Zr ≈ 3:7; 0.1-0.3	$\rho \approx 500 \mu\Omega$ cm TCR ≈ 0 ppm °C^{-1}	
Mixture of TiN, AlN with Ti:Al ≈ 54:46; 0.1-0.3	$\rho \approx 600 \mu\Omega$ cm TCR ≈ 0 ppm °C^{-1}	Cermet-like structure of conductive TiN and insulating AlN
Mixture of Ti-Zr-N, AlN with Ti:Zr:Al ≈ 1:2:1; 0.1-0.3	$\rho \approx 7800 \mu\Omega$ cm TCR ≈ -200 ppm °C^{-1}	Cermet-like structure of conductive Ti-Zr-N and insulating AlN

[d] For pure argon $K \approx 1.5$.
[e] For pure argon $K \approx 0.7$.
[f] For pure argon $K \approx 0.6$.

Table 5.56 continued.

5.10 REFERENCES

1. Foster, N.F., Coquin, G.A., Rozgonyi, G.A., and Vannatta, F.A., IEEE Trans. Sonics Ultrason., SU-15: 28 (1968).

2. Matsumoto, S., Oyo Butsuri, 49: 43 (1980).

3. Wasa, K., and Hayakawa, S., Jpn. J. Appl. Phys., 12: 408 (1973).

4. Nowicki, R.S., J. Vac. Sci. Technol., 14: 127 (1977).

5. Izama, T,, Mori, H., Murakami, Y., and Shimizu, N., Appl. Phys. Lett., 38: 483 (1981).

6. Hartsough, L.D., and McLeod, P.S., J. Vac. Sci.Technol., 14: 123 (1977).

7. Khuri-Yakub, B.T., Kino, G.S., and Galle, P., J. Appl. Phys., 46: 475 (1976).

8. Ohji, K., Tohda, T., Wasa, K., and Hayakawa, S., J. Appl. Phys., 47: 1726 (1976).

9. Shiosaki,T., IEEE Ultrasonics Symp. Proc., (1978) p.100.

10. Mitsuyu, T., Ono, S., and Wasa, K., J. Appl. Phys., 51: 2464 (1980).

11. Mitsuyu, T., Ono, S., and Wasa, K., Proc. Symp. Fundamentals and Appl. of Ultrasonic Electronics, Tokyo, p. 55 (1980).

12. Mitsuyu, T., Wasa, K., and Hayakawa, S., J.Electrochem.Soc., 123: 94 (1976).

13. Takei, W.J., Formigoni, N.P., and Francombe, M.H., J.Vac. Sci. Technol., 7: 442 (1969).

14. Sugibuchi, K., Kurogi, Y., and Endo, N., J Appl. Phys., 46: 2877 (1975).

15. Mitsuyu, T., Wasa, K., and Hayakawa, S., J. Crystal Growth, 41: 151 (1977).

16. Mitsuyu, T., Wasa, K., and Hayakawa, S., J. Appl. Phys., 47: 2901 (1976).

17. Payne, D.A., and Mukerjee, J.L., Appl. Phys. Lett., 29: 748 (1976).

18. Buchanan, M., Webb, J.B., and Williams, D.F., Appl. Phys. Lett., 37: 213 (1980).

19. Hori, M., Adachi, M., Shiosaki, T., and Kawabata, A., Fall Meeting of Oyo Butsuri-Gakukai, paper 3p-F-1O (1978).

20. Mitsuyu, T., and Wasa, K., Jpn. J. Appl. Phys., 20: L48 (1981).

21. Takada, S., Ohnishi, M., Hayakawa, H., and Mikoshiba, N., Appl. Phys. Lett., 24: 490 (1974).

22. Miyazawa, S., Fushimi, S., and Kondo, S., Appl. Phys. Lett., 26: 8 (1975). (1975) 8.

23. Kusao, K., Wasa, K., and Hayakawa, S., Jpn. J. Appl.Phys., 7: 437 (1968).

24. Hamada, H., Morooka, H., and Hirai, H., Spring Meeting of Oyo Butsuri-Gakukai, paper 27p-H-6 (1979).

25. Oikawa, M., and Toda, K., Appl. Phys. Lett., 29: 491 (1976).

26. Okada, A., J. Appl. Phys., 48: 2905 (1977).

27. Usuki, T., Nakagawa, T., Okuyama, M., Kariya, T., and Hamakawa, Y., Fall Meeting of Oyo Butsuri-Gakukai, paper 2p-F-1O (1978).

28. Matsui, Y., Okuyama, M., and Hamakawa, Y., 1st Meeting Ferroelectric Mat.. and Appl., Kyoto, paper 17-T-3 (1977).

29. Ishida, M., Tsuji, S., Kimura, K., Matsunami, H., and Tanaka, T., J. Cryst. Growth, 45: 393 (1978).

30. Sato, S., Oyo Butsuri, 47: 656 (1978).

31. Wasa, K., and Hayakawa, S., Thin Solid Films, 1O: 367 (1972).

32. Takeda, F., and Hata, T., Jpn. J. Appl. Phys., 19: 1001 (1980).

33. Shiosaki, T., Yamamoto, T., Oda, T., and Kawabata, A., Appl. Phys. Lett., 36: 643 (1980).

34. Shuskus, A.J., Reeder, T.M., and Paradis, E.L., Appl. Phys. Lett., 24: 155 (1974).

35. Liu, J.K., Lakin, K.M., and Wang, K.L., J. Appl. Phys., 46: 3703 (1975).

36. Agarwal, V., Vankar, V.D., and Chopra, K.L., J. Vac. Sci. Technol. A6: 2361 (1988).

37. Chopra, K.L., Agarwal, V., Vankar, V.D., Deshpandey, C.V., and Bunshah, R.F., Thin Solid Films, 126: 307 (1985).

38. Mito, H., and Horiguchi, S., Ionics, 36: 10 (1978).

39. Wasa, K., and Hayakwa, S., Microelectron Reliab., 6: 213 (1967).

40. Srivastava, P.K., Rao, T.V., Vankar, V.D., and Chopra, K.L., J. Vac. Sci. Technol. A2: 1261 (1984), Srivastava, P.K., Vankar, V.D., and Chopra, K.L., J. Vac. Sci. Technol. A3: 2129 (1985), ibid. A4: 2819 (1986), Nov/Dec (1985) 2129, ibid. A4(6), Nov/Dec (1986) 2819, Srivastava, P.K., Vankar, V.D., and Chopra, K.L., Bull. Mater. Sci., 8: 379 (1986)., Srivastava, P.K. Vankar, V.D., and Chopra, K.L., Thin Solid Films 166: 107 (1988).

41. Wasa, K., Nagai, T., and Hayakawa, S., Thin Solid Films, 31: 235 (1976).

42. Agarwal, V., Vankar, V.D., and Chopra, K.L., J. Vac. Sci. Technol. A6: 2341 (1988).

43. Wasa, K., Tohda, T., Kasahara, Y., and Hayakawa, S., Rev. Sci. Instr., 50: 1086 (1979).

44. Murayama, Y., and Takao, T., Thin Solid Films, 40: 309 (1977).

45. Ono, H., Nishino, S., and Matsunami, H., IECE, Japan, Tech. Reps., SSD 80-80: 125 (1981).

46. Stringfellow, G.B., Stall, R., and Koschel, W., Appl. Phys. Lett., 38: 156 (1981).

47. Barnett, S.A., Bajor, G., and Greene, J.E., Appl.Phys.Lett., 37: 734 (1980).

48. Chang, C.A., Ludeke, R., Chang, L.L., and Esaki, L., Appl. Phys. Lett., 31: 759 (1977).

49. Greene, J.E., Wickersham, C.E., and Zilko, J.L., Thin Solid Films, 32: 51 (1976).

50. Wu, C.T., Kampwirth, R.T., and Hafstrom, J.W., J. Vac. Sci. Technol., 14: 134 (1977).

51. Ihara, H., Kimura, Y., Okuyama, H., and Gonda, S., IEEE Trans. Magn. MAG-19: 938 (1983).

52. Kawamura, H., and Tachikawa, K., Phys. Lett., 5OA: 29 (1974).

53. Gavaler, J.R., Greggi, J., Wilmer, R., and Ekin, J.W., IEEE Trans. Magn. MAG-19: 418 (1983).

54. Smith, W.D., Lin, R.Y., Coppola, J.A., and Economy, J., IEEE Trans. Magn. MAG-11: 182 (1975).

55. Dietrich, M., Dustmann, C.H., Scmaderer, F., and Whahl, G., IEEE Trans. Magn. MAG-19: 406 (1983).

56. Alterovitz, S.A., Woolam, J.A., Kammerdiner, L., and Luo, H.L., Appl. Phys. Lett., 33: 264 (1978).

57. Hamasaki, K., Inoue, T., Yamashita, T., and Komata, T., Appl. Phys. Lett., 41: 667 (1982).

58. Suzuki, M., Enomoto, Y., Murakami, T., and Inamura, T., J. Appl. Phys., 53: 1622 (1982).

59. Murakami, T., and Suzuki, M., Jpn. J. Appl. Phys., 24: Suppl. 24-2, p. 323 (1985).

60. Terada, N., Ihara, H., Hirabayashi, M., Senzaki, K., Kimura, Y., Murata, K., and Tokumoto, M., Jpn. J. Appl. Phys., 26: L508 (1987).

61. Suzuki, M., and Murakami, T., Jpn. J. Appl. Phys., 26: L524 (1987).

62. Nagata, S., Kawasaki, M., Funabashi, M., Fueki, K., and Koinuma, H., Jpn. J. Appl. Phys., 26: L410 (1987).

63. Adachi, H., Setsune, K., and Wasa, K., Phys. Rev.B, 35: 8824 (1987).

64. Naito, M., Smith, D.P.E., Kirk, M.D., Oh, B., Hahn, M.R., Char, K., Mitzi, D.B., Sun, J.Z., Webb, D.J., Beasley, M.R., Fischer, O., Geballe, T.H., Hammond, R.H., Kapitulnik, A., and Quate, C.F., Phys. Rev. B, 35: 7228 (1987).

65. Chaudhari, P., Koch, R.H., Laibowitz, R.B., McGuire, T.R., and Gambino, R.J., Phys. Rev. Lett., 58: 2684 (1987).

66. Oh, B., Nait, M., Arnason, S., Rosenthal, P., Barton, R., Beasley, M.R., Geballe, T.H., Hamond, R.H., and Kapitulnik, A., Appl. Phys. Lett., 51: 852 (1987).

67. Enomoto, Y., Murakami, T., Suzuki, M., and Moriwaki, K., Jpn. J. Appl. Phys., 26: L1248 (1987).

68. Adachi, H., Setsune, K., Mitsuyu, T., Hirochi, K., Ichikama, Y., Kamada, T., and Wasa, K., Jpn. J. Appl. Phys., 26: L709 (1987).

69. Adachi, H., Setsune, K., and Wasa, K., Proc. of 18th Int. Conf. on Low Temp. Physics, Kyoto, DM34 (1987).

70. Dijkkamp, D., and Venkatesan, T., Appl. Phys. Lett., 51: 619 (1987).

71. Tsaur, B-Y., Dilorio, M.S., and Strauss, A.J., Appl.Phys. Lett., 51: 858 (1987).

72. Adachi, H., Setsune, K., Hirochi, K., Kamada, T., and Wasa, K., Proc. Int. Conf. on High-Temp. Superconductors and Mat. and Mech. of Superconductivity, Interlaken, Switzerland (1988)., Ichikawa, Y., Adachi, H., Hirochi, K., Setsune, K., Hatta, S. and Wasa, K., Phys. Rev. B, 36: (1988).

73. Kang, J.H., Kampwirth, R.T., and Gray, K.E., Appl. Phys. Lett., 52: 2080 (1988), Kang, J.H., Kampwirth, R.T., Gray, K.E., Marsh, S., and Huff, E.A., Phys. Lett. 128: 102 (1988).

74. Kuroda, K., Mukaida, M., Yamamoto, M., and Miyazawa, S., Jpn. J. Appl. Phys. 27: L625 (1988).

75. Rice, C.E., Levi, A.F.J., Fleming, R.M., Marsh, P., Baldwin, K.W., Anzlower, M., White, A.E., Short, K.T., Nakahara, S., and Stormer, H.L., Appl. Phys. Lett. 52: 1828 (1988).

76. Kanai, M., Kawai, T., Kawai, M., and Kawai, S., Jpn. J. Appl. Phys. 27: L1293 (1988).

77. Lolentz, R.D., and Sexton, J.H., Appl. Phys. Lett. 53: 1654 (1988).

78. Adachi, H., Kohiki, S., Setsune, K., Mitsuyu, T., and Wasa, K., Jpn. J. Appl. Phys. 27: L1883 (1988).

79. Yamane, H., Kurosawa, H., Hirai, T., Iwasaki, H., Kobayashi, N., and Muto, Y., Jpn. J. Appl. Phys. 27: L1495 (1988).

80. Ginley, D.S., Kwak, J.F., Hellmer, R.P., Baughman, R.J., Venturini, E.L., and Morosin, B., Appl. Phys. Lett. 53: 406 (1988).

81. Kang, J.H., Kampwirth, R.S., and Gray, K.E., Phys. Lett. 53: (1988).

82. Lee, W.Y., Lee, V.Y., Salem, J., Huang, T.C., Savoy, R., Bullock, D.C., and Parkin, S.S.P., Appl. Phys. Lett. 53: 329 (1988).

83. Hong, M., Liou, S.H., Bacon, D.D., Grader, G.S., Kwo, J., Kortan, A.R., and Davison, B.A., Appl. Phys. Lett. 53: 2102 (1988).

84. Adachi, H., Wasa, K., Ichikawa, Y., Hirochi, K., and Setsune, K., J. Cryst. Growth 91: 352 (1988).

85. Ichikawa, Y., Adachi, H., Setsune, K., Hatta, S., Hirochi, K., and Wasa, K., Appl. Phys. Lett. 53: 919 (1988).

86. Wasa, K., Adachi, H., Ichikawa, Y., Hirochi, K., and Setsune, K., Proc. of Conf. on the Sci. & Technol. of Thin Film Superconductors, Colorado Springs, CO (Nov. 1988).

87. Kitabake M., and Wasa, K., J. Appl. Phys., 58: 1693 (1985).

88. Muller, K-H. J. Appl. Phys., 59: 2803 (1986).

89. Muller, K-H. J. Vac. Sci. Technol., A5: 2161 (1987).

90. Chubachi, N., Oyo Butsuri, 46: 663 (1973).

91. Galli, C., and Coker, J.E., Appl. Phys. Lett., 16: 439 (1970).

92. Machida, K., Shibutani, M., Murayama, Y., and Matsumoto, M., Trans. IECE, 62: 358 (1979).

93. Ohji, K., Yamazaki, O., Wasa, K., and Hayakawa, S., J.Vac. Sci. Technol., 15: 1601 (1978).

94. Wasa, K., and Hayakawa, S., Oyo Butsuri, 6: 580 (1981).

95. Shiosaki, T., Oonishi, M., Oonishi, S., and Kawabata, A., Proc. 1st Meeting Ferroelectric Materials and their Applications, Kyoto, Japan, 7 (1977).

96. Westwood, W.D., and Ingrey, S.J., Wave Electronics, 1: 139 (1974/1975).

97. Hada, T., Wasa, K., and Hayakawa, S., Thin Solid Films, 7: 135 (1971).

98. Foster, N.F., J. Vac. Sci. Technol., 6: 111 (1969).

99. Chubachi, N., Proc. IEEE, 64: 772 (1976).

100. Manabe, Y., Mitsuyu, T., Yamazaki, O., and Wasa, K., Proc. 1988 Spring Meeting of Jpn. Appl. Phys., 28pS-8/II.

101. Mitsuyu, T., Ono, S., and Wasa, K., J. Appl. Phys., 51: 2464 (1980).

102. Mitsuyu, T., Yamazaki, O., Ohji, K., and Wasa, K., J. Cryst. Growth, 42: 233 (1982).

103. Tiku, S.K., Lau, C.K., and Lakin, K.M., Appl. Phys. Lett., 36: 318 (1980). (1980) 318-320.

104. Shiozaki, T., Ohnishi, S., Hirokawa, Y., and Kawabata, A. Appl. Phys. Lett., 33: 406 (1978).

105. Ohji, K., Tohda, T., Wasa, K., and Hayakawa, S., J. Appl. Phys., 47: 1726 (1976).

106. Wasa, K., Hayakawa, S., and Hada, T., IEEE Trans.Sonics Ultrason., SU-21: 298 (1974).

107. Hickernell, F.S., J. Solid State Chemistry, 12: 225 (1975).

108. Kino, G.S., and Wagers, R.S., J. Appl. Phys., 44: 1480 (1973).

109. Denburg, D.L., IEEE Trans. Sonics & Ultrason., SU-18: 31 (1971).

110. Shiosaki, T., Yasumoto, Y., and Kawabata, A., Proc.6th Conf. Solid Devices, Tokyo (1974), Jpn. J. Appl. Phys., 44: suppl. 11 (1975): suppl., p.115.

111. Ohji, K., and Wasa, K., Proc. 1981 Meeting of IEEE Japan, 500.

112. Kushibiki, J., and Chubachi, N., IEEE Trans. Sonics & Ultrason., SU-32: 189 (1985).

113. White, R.M., Proc. IEEE, 58: 1238 (1970).

114. Hays, R.M., and Hartmann, C.S, Proc. IEEE, 64: 652 (1976).

115. Yamazaki, O., Mitsuyu, T., and Wasa, K., IEEE Trans. Sonics and Ultrason., SU-27: 369 (1980).

116. Setsune, K., Mitsuyu, T., Yamazaki, O., and Wasa, K., 1983 Ultrasonic Sympoium Proc. Atlanta, p.467, Suhara, T., Shiono, T., Nishihara, H., and Koyama, J., IEEE J. Lightwave Tech., LT-1: 624 (1983).

117. Wasa, K., and Hayakawa, S., Jpn. J. Appl. Phys., 10: 1732 (1971).

118. Lou, L.F., J. Appl. Phys., 50: 555 (1979).

119. Lenzo, P.V., et al., J. Appl. Opt., 5: 1688 (1966).

120. Venturini, E.V., et al., J. Appl. Phys., 40: 1622 (1969).

121. Gattow, Z., et al., Z.Anorgo, Allg. Chem., 318: 176 (1962). Ballman, A.A., J. Cryst. Growth, 1: 37 (1967).

122. Feldman, C., Rev. Sci. Instrum., 26: 463 (1955).

123. Wasa, K., and Hayakawa, S., Thin Solid Films, 52: 31 (1978).

124. Kusao, K., Wasa, K., and Hayakawa, S., Jpn. J.Appl.Phys., 7: 437 (1968).

125. Kitabatake, M., and Wasa, K., Jpn. J. Appl. Phys., Suppl. 24: 33 (1985).

126. Adachi, H., Mitsuyu, T., Yamazaki, O., and Wasa, K., Jpn. J. Appl. Phys., Suppl. 24: 13 (1985).

127. Adachi, H., Mitsuyu, T., Yamazaki, O., and Wasa, K., J. Appl. Phys. 60: 736 (1986).

128. Kushida, K., and Takeuchi, H., Appl. Phys. Lett., 50: 1800 (1987).

129. Iijima, K., Kawabata, S., and Ueda, I., Proc. of The 3rd Sensor Symp., Tsukuba, 133 (1983).

130. Adachi, H., Mitsuyu, T., Yamazaki, O., and Wasa, K., Jpn. J. Appl. Phys., Suppl., 26: 15 (1987).

131. Farnell, G.W., Cermak, I.A., Silvester, P., and Wong, S.K., IEEE Trans. Sonics Ultrason. SU-17: 188 (1970).

132. Adachi, H., Kawaguchi, T., Kitabatake, M., and Wasa, K., Jpn. J. Appl. Phys., Suppl. 22-2: 11 (1983).

133. Keizer, K., and Burggraaf, A.J., Ferroelectrics 14: 671 (1976).

134. Smith, W.R., Gerard, H.M., Collins, J.H., Reeder, T.M., and Shaw, H.J., IEEE Trans. MTT, MTT-17: 856 (1969).

135. Bechmann, R., J. Acoust. Soc. Am. 28: 347 (1956).

136. Gavrilyachenko, V.G., and Fesenko, E.G., Sov. Phys. Crystallogr. 16: 549 (1971).

137. Wasa, K., Yamazaki, O., Adachi, H., Kawaguchi, T., and Setsune, K., J. Lightwave Technol., LT-2: 710 (1984).

138. Tsai, C.S., Kim, B., and El-akkari, F.R., IEEE J.Quantum Electron., QE-14: 513 (1978).

139. Adachi, H., and K.Wasa, K., ISIAT'83 & IPAT'83, Kyoto, 951 (1983), Adachi, H., Mitsuyu, T., and Wasa, K., Jpn. J. Appl. Phys. suppl., 24-1: 121 (1985).

140. Adachi, H., Mitsuyu, T., Yamazaki, O., and Wasa, K., Jpn.J. Appl. Phys., Suppl. 24-2: 287 (1985).

141. Zang, D.Y., and Tsai, C.S., Applied Optics, 25: 2264 (1986).

142. Bednorz, J.G., and Muller, K.A., Z. Phys. B64: 189 (1986).

143. Caponell, D.W., Hinks, D.G., Jorgensen, J.D., and Zhang, K., Appl. Phys. Lett., 50: 543 (1987).

144. Wu, M.K., Ashburn, J.R., Torng, C.J., Hor, P.H., Meng, R.L., Gao, L., Huang, Z.H., Wang, Y.Q., and Chu, C.W., Phys.Rev. Lett. 58: 908 (1987).

145. Maeda, I., Tanaka, Y., Fukmori, M., and Asano, T., Jpn.J. Appl. Phys., 27: L209 (1988).

146. Sheng, Z.Z., and Hermann, A.M., Nature, 332: 138 (1988).

147. Sleight, A.W., Gillson, J.L., and Bierstedt, F.E., Solid State Commun. 18: 27 (1975).

148. Poole Jr, C.P., Datta, T., and Farach, H.A., Copper Oxide Superconductors, New York, John Wiley & Sons (1988).

149. Fueki, F., Kitazawa, K., Kishio, K., and Hasegawa, T., Proc. Srinagar Workshop on High Temperature Superconductivity, Srinagar (A.K. Gupta, S.K. Joshi, C.N.R. Rao, eds.) World Scientific, Singapore (1988) p.119.

150. Chaudhari, P., Collins, R.T., Freitas, P., Gambino, R.J., Kirtley, J.R., Koch, R.H., Laibowitz, R.B., LeGoues, F.K., McGuire, T.R., Penney, T., Schlesinger, Z., and Segmuller, A.M., Phys. Rev. B 36: 8903 (1987).

151. Adachi, H., Setsune, K., Hirochi, K., Kamada, T., Wasa K., Proc. Int. Conf. on High-Temp. Superconductors and Materials and Mechanisms of Superconductivity, E12, Interlaken, Switzerland (1988).

152. Hammond, R.H., Naito, M., Oh, B., Hahn, M., Rosenthal, P.,. Marshall, A., Missert, N., Beasley, M.R., Kapitulnik, A., and Geballe, T.H., Extended Abstracts for MRS Symposium on High Temperature Superconductors, Anaheim, California (1987).

153. Hirochi, K., Adachi, H., Setsune, K., Yamazaki, O., and Wasa, K., Jpn. J. Appl. Phys., 26: L1837 (1987).

154. Nastasi, M., Arendt, P.N., Tesmer, J.R., Maggiore, C.J., Cordi, R.C., Bish, D.L., Thompson, J.D., Cheong, S.-W., Bordes, N., Smith, J.F., and Raistrick, I.D., J. Mater. Res. 2: 726 (1987).

155. Hatta, S., Higashino, H., Hirochi, K., Adachi, H., and Wasa, K., Appl. Phys. Lett., 53: 148 (1988).

156. Tsaur, B.-Y., Dilorio, M.S., and Strauss, A.J., Appl. Phys. Lett., 51: 858 (1987).

157. Narayan, J., Biunno, N., Singh, R., Holland, O.W., and Auciello, O., Appl. Phys. Lett., 51: 1845 (1987).

158. Berry, A.D., Gaskill, D.K., Holm, R.T., Cukauskas, E.J., Kaplan, R., Henry, R.L., Appl. Phys. Lett., 52: 1743 (1988).

159. Adachi, H., Kohiki, S., Setsune, K., Mitsuyu, T., and Wasa, K., Jpn. J. Appl. Phys., 27: L1883 (1988).

160. Higashino, H., Enokihara, A., Mizuno, K., Mitsuyu, T., Setsune, K., and Wasa, K., 5th International Workshop on Future Electron Devices High-Temperature Super-conducting Electron Devices, (FED HiTcSc-ED WORKSHOP), Miyagi-Zao, p.267 (1988).

161. Enokihara, A., Higashino, H., Setsune, K., Mitsuyu, T., and Wasa, K., Jpn. J. Appl. Phys., 27: L1521 (1988).

162. Hirao, T., Kamada, T., Kitagawa, M., Hayashi, S., Miyauchi, M., Setsune, K., and Wasa, K., J. Vac. Sci. Technol. (submitted).

163. Kitabatake, M., Mitsuyu, T., and Wasa, K., J.Non-Cryst. Solids 53: 1 (1982).

164. Oikawa, M., and Toda, K., Appl. Phys. Lett. 29: 491 (1976).

165. Dijkkamp, D., Venkatesan, T., Wu, X.D., Shaheen, S.A., Jisrawi, N., Min-Lee, Y.H., McLean, W.L., and Croft, M., Appl. Phys. Lett. 51: 619 (1987).

166. Fukami, T., and Sakuma, T., Jpn. J. Appl. Phys. 2O: 1599 (1981).

167. Nakagama, T., Yamaguchi, J., Okuyama, M., and Hamakawa, Y., Jpn. J. Appl. Phys. 21: L655 (1982).

168. Wehner, G.K., Kim, Y.H., Kim, D.H., and Goldman, A.M., Appl. Phys. Lett. 52: 1187 (1988).

169. Ihara, M., and Kimura, T., 5th International Workshop on Future Electron Devices High-temperature Superconducting Electron Devices, (FED HiTcSc-ED WORK-SHOP), Miyagi-Zao, p.137 (1988).

170. Wasa, K., Adachi, H., Ichikawa, Y., Setsune, K., and Hirochi, K., Proc. International Workshop on High T_c Superconductors, Srinagar (1988).

171. Gurvitch M., and Fiory, A.T., Appl. Phys. Lett., 51: 1027 (1987).

172. Ichikawa, Y., Adachi, H., Mitsuyu, T., and Wasa, K., Jpn.J. Appl. Phys., 27: L381 (1988).

173. Takayama-Muromachi, E., Uchida, Y., Yukino, K., Tanaka, T., and Kato, K., Jpn. J. Appl. Phys. 26: L665 (1987).

174. Jorgensen, J.D., Beno, M.A., Hinks, D.G., Solderholm, L., Volin, K.J., Hitterman, R.L., Grace, J.D., and Shuller, I.K., Phys. Rev. B 36: 3608 (1987).

175. Hayashi, S., Kamada, T., Setsune, K., Hirao, T., Wasa, K., and Matsuda, A., Jpn. J. Appl. Phys. 27: L1257 (1988).

176. Ichikawa, Y., Kitabatake, M., Kohiki, S., Adachi, H., Hatta, S., Setsune, K., and Wasa, K., Jpn. J. Appl. Phys. (submitted).

177. Kambe, S., and Kawai, M., Jpn. J. Appl. Phys. 27: L2342 (1988).

178. Li, H.C., Linker, G., Ratzel, F., Smithey, R., and Geek, J., Appl. Phys. Lett. 52: 1098 (1988).

179. Wu, X.D., Inam, A., Venkatesan, T., Chang, C.C., Chase, F.W., Barboux, P., Tarascon, J.M., and Wilkens, B., Appl. Phys. Lett. 52: 754 (1988).

180. Terashima, T., Iijima, K., Yamamoto, K., Bando, Y., and Mazaki, H., Jpn. J. Appl. Phys. 27: L91 (1988).

181. Hatta, S., and Wasa, K., Proc. 55th Magnetics Symposium, Tokyo, p.7 (1988).

182. Setsune, K., Adachi, H., Kamada, T., Hirochi, K., Ichikawa, Y., and Wasa, K., in Extended Abstracts, 8th Int. Symp. on Plasma Chem., Tokyo, p.2335 (1987).

183. Chaudhari, P., Koch, R.H., Laibowitz, R.B., McGuire, T.R., and Gambino, R.J., Phys. Rev. Lett. 58: 2684 (1987).

184. Adachi, H., Setsune, K., and Wasa, K., Jpn. J. Appl. Phys., Suppl. 26-3: 1139 (1987).

185. Wasa, K., Kitabatake, M., Adachi, H., Setsune, K., and Hirochi, K., AIP Conf. Proc. 165: 37 (1988).

186. Kamada, T., Setsune, K., Hirao, T., and Wasa, K., Appl. Phys. Lett., 52: 1726 (1988).

187. Takayama-Muromachi, E., Uchida, Y., Ono, A., Izmi, F., Onoda, M., Matsui, Y., Kosuda, K., Takekawa, S., and Kato, K., Jpn. J. Appl. Phys., 27: L365 (1988).

188. Adachi, H., Wasa, K., Ichikawa, Y., Hirochi, K., and Setsune, K., J. Cryst. Growth 91: 352 (1988).

189. Ichikawa, Y., Adachi, H., Setsune, K., Hirochi, K., Hatta, S., and Wasa, K., Appl. Phys. Lett. 53: 919 (1988).

190. Satoh, T., Yoshitake, T., Kubo, Y., and Igarashi, H., Appl. Phys. Lett. 53: 1213 (1988).

191. Hatta, S., Ichikawa, Y., Hirochi, K., Adachi, H., and Wasa, K., Jpn. J. Appl. Phys., 27: L855 (1988).

192. Setsune, K., Hirochi, K., Adachi, H., Ichikawa, Y., and Wasa, K., Appl. Phys. Lett. 15: 600 (1988).

193. Ogale, S.B., Dijkkamp, D., Venkatesan, T., Wu, X.D., and Inam, A., Phys. Rev. B 36: 7210 (1987).

194. Wasa, K., Adachi, H., Ichikawa, Y., Hirochi, K., and Setsune, K., Proc. ISS'88, Nagoya (1988).

195. Kohiki, S., Hirochi, K., Adachi, H., Setsune, K., and Wasa, K., Phys. Rev. B, 38: 9201 (1986), ibid, 39: 4695 (1988).

196. Adachi, H., Kohiki, S., Setsune, K., Mitsuyu, T., and Wasa, K., Jpn. J. Appl. Phys., 27: L1883 (1988).

197. Hatta, S., Hirochi, K., Adachi, H., Kamada, T., Ichikawa, Y., Setsune, K., and Wasa, K., Jpn. J. Appl. Phys., 27: 1646 (1988).

198. Bean, C.P., Rev. Mod. Phys. 36: 31 (1964).

199. Dinger, T.R., Worthington, T.K., Gallagher, W.J., and Sandstrom, R.L., Phys. Rev. Lett. 58: 2687 (1987), Van Dover, R.B., Schneemeyer, L.F., Gyorgy, E.M., and Waszczak, J.V., Appl. Phys. Lett. 52: 1910 (1988).

200. Tokura, Y., Takagi, H., and Uchida, S., Nature, 337: 345 (1989). Hayashi, S., Adachi, H., Setsune, K., Hirao, T., and Wasa, K., Jpn. J. Appl. Phys. (submitted).

201. Katsube, T., and Katsube, Y., Oyo Butsuri, 49: 2 (1980).

202. Fraser, D.B., and Cook, H.D., J. Electrochem. Soc., 119: 1368 (1972).

203. Ohata, Y., and Yoshida, S., Oyo Butsuri, 46: 43 (1977).

204. Buchanan, M., Webb, J.B., and Williams, D.F., Appl. Phys. Lett., 37: 213 (1980).

205. McLean, D.A., Schwartz, N., and Tidd, E.D., IEEE Int. Conv. Record, 12: pt-9, 128 (1964).

206. Wasa, K., and Hayakawa, S., Thin Solid Films, 52: 31 (1978).

207. Wasa, K., and Hayakawa, S., Thin Solid Films, 1O: 367 (1972).

208. Wasa, K., and Hayakama, S., U. S. Pat. 3803057 (1974).

209. Stein, H.J., Picraux, S.T., and Holloway, P.H., IEEE Trans. Electron Devices 25: 587 (1978).

210. Fujita, S., Toyoshima, H., Ohishi, T., and Sasaki, A., Jpn. J. Appl. Phys. 23: L268 (1984).

211. Chow, R., Lanford, W.A., Kem-Ming, W., and Rosler, R.S., J. Appl. Phys. 53: 5630 (1982).

212. Kitabatake, M., and Wasa, K., Appl. Phys. Lett. 49: 927 (1986).

213. ibid.

214. Khan, I.H., and Summergrad, R.N., Appl. Phys. Lett. 11: 12 (1967). 12.

215. Yoshihara, H., Mori, H., Kiuchi, M., and Kadota, T., Jpn. J. Appl. Phys. 17: 1693 (1978).

216. Learn, A.J., and Haq, K.E., Appl. Phys. Lett. 17: 26 (1970).

217. Wasa, K., Nagai, T., and Hayakawa, S., Thin Solid Films, 31: 235 (1976).

218. Murayama, Y., and Takao, T., Thin Solid Films, 40: 309 (1977).

219. Lipson, H.G., in Silicon Carbide, (J.R. O'Connor and J. Smiltens, eds.) 371, Pergamon, London, (1960).

220. Wasa, K., Tohda, T., Kasahara, Y., and Hayakawa, S., Rev. Sci. Instr. 50: 1084 (1979).

221. Carborundum Company, Cat. No. 32602.

222. Wasa, K., Koubunshi, 35: 138 (1986).

223. Aisenberg, S., and Chabot, R., J. Appl. Phys., 42: 2953 (1971).

224. Wasa, K., and Hayakawa, S., Jpn. Pat. (S51-84840).

225. Weissmantel, C., Bewilogua, K., Dietrich, D., Hinneberg, H.-J., Klose, S., Nowick, W., and Reisse, G., Thin Solid Films, 72: 19 (1980).

226. Kitabatake, M., and Wasa, K., J. Appl. Phys., 58: 1693 (1985).

227. Calow, J.T., Kirk, D.L., Owen, S.T.J., and Webb, P.W., Radio and Elect. Eng. 41: 243 (1971).

228. Tsuya, N., and Arai, K.I., Jpn. J. Appl. Phys., 18: 461 (1979).

229. Kitabatake, M., Mitsuyu, T., and Wasa, K., Oyo Butsuri, 6: 568 (1985).

230. Kitabatake, M., Mitsuyu, T., and Wasa, K., J. Appl. Phys., 56: (1985)

231. Kitabatake, M., Mitsuyu, T., and Wasa, K., J. Non-Cryst. Solids, 53: 1 (1982).

232. Kinbara, A., and Uosmi, K., Oyo Butsuri, 45: 1165 (1976).

233. Stroud, P.T., Thin Solid Films, 11: 1 (1972).

234. Lines, M.E., Phys. Rev. B15: 388 (1977).

235. Glass, A.M., Lines, M.E., Nassau, K., and Shiever, J.W., Appl. Phys. Lett. 31: 249 (1977).

236. Glass, A.M., Nassau, K., and Negran, T.J., J. Appl. Phys. 49: 4808 (1978).

237. Takashige, M., Nakamura, T., Tsuya, N., Arai, K., Ozawa, H., and Uno, R., Jpn. J. Appl. Phys., 19: L555 (1980).

238. Mitsuyu, T., and Wasa, K., Jpn. J. Appl. Phys., 20: L48 (1981).

239. Nassau, K., Levinstein, H.J., and Loiacono, G.M., J. Phys. Chem. Solids, 27: 989 (1966).

240. Kitabatake, M., and Wasa, K., J. Non-Crystalline Solids, 58: 1 (1982). Spring Meeting Phys. Soc. Japan. paper 2a-J-12.

241. Takashige, M., Cho, T., Nakamura, T., Aikawa, Y., paper 2a-J-12, Spring Meeting Phys., Soc., Japan (1982).

242. Tohda, T., Wasa, K., and Hayakawa, S., J. Electrochem. Soc., 127: 44 (1980).

243. Lipson, H.G., in Silicon Carbide, (J.R. O'Connor, J. Smiltens, eds.) p. 371, Pergammon Press, London (1960).

244. Moustakas, T. D., and Friedman, R., Appl. Phys. Lett., 40: 515 (1982).

245. Esaki, L. Proc. 6th Int. Vac. Congr., Kyoto (1974), Jpn. J. Appl. Phys., 13: Suppl. 2, Pt. 1, 821 (1974).

246. Pessa, M.V., Huttunen, P., and Herman, M.A., J. Appl., Phys., 54: 6047 (1983).

247. Wasa, K., Kitabatake, M., Adachi, H., Ichikawa, Y., Hirochi, K., and Setsune, K., Thin Solid Films, 181: 199 (1989).

248. Greene, J.E., Wickersham, C.E., and Zilko, J.L., J. Appl. Phys., 47: 2289 (1976).

249. Eltoukhy, A.H., Zilko, J.L., Wickersham, C.E., and Greene, J.E., Appl. Phys. Lett., 31: 156 (1977).

250. Kitabatake, M., Ph.D. Thesis (1988).

251. Hada, T., Hayakawa, S., and Wasa, K., Jpn. J. Appl. Phys., 9: 1078 (1970).

252. Chopra, K.L., Thin Film Phenomena, New York, McGraw-Hill (1969).

253. Wasa, K., and Hayakawa, S., Thin Solid Films, 52: 31 (1978).

6

MICROFABRICATION BY SPUTTERING

Atoms of a solid surface are removed under irradiation of energetic ions. This phenomena is called "sputter etching." The sputter etching process is governed by collisions between the irradiated atoms and the surface atoms of the solid target. The interaction depth for sputter etching is around 100 Å beneath the surface. The sputter etching process is useful for submicron fabrication.

When the surface of a solid is bombarded by argon ions the adsorbed gases are removed and a clean surface is produced. Thinning of specimens is achieved by argon ion bombardment for fabrication of test pieces for electron microscope analysis. Bombardments by chemically reactive ions achieve surface etching through chemical sputtering. Sputter etching processes are useful for making masking patterns for LSI.

Ion beam sputtering and/or diode sputtering are used for surface fabrications. In this chapter, microfabrication processes are described in relation to fabrication of thin film electronic devices. Detailed descriptions for semiconducting LSI have been reviewed in several books (1-3).

6.1 ION BEAM SPUTTER ETCHING

The most useful system for the sputter etching is the ion beam sputtering system as shown in previous chapter. The sputtering chamber is separated from the ion source and the surface of the specimens is not revealed to the plasma generated in the ion source. The sputtering chamber is generally kept below 1×10^{-4} Torr during the sputter etching.

Several ion sources are proposed including a hot cathode ion source, cold cathode ion source and a plasmatron ion source (2,3). Typical constructions are shown in Fig. 6.1.

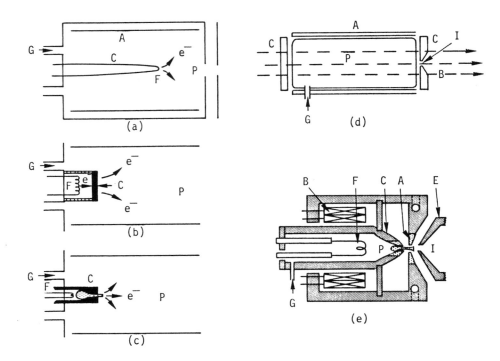

Figure 6.1: Typical constructions of ion sources. C; cathode A; anode P; plasma B; magnetic field G; gas inlet I; window E; accelerating electrode F; filament.

The hot cathode type is so called "Kaufman ion source" which is widely used for the sputter etching system. In the Kaufman ion source the tungsten and/or oxide cathode are used for the hot cathode. The reactive gas often damages the hot cathode. A hollow cathode ion source where the hot cathode is mounted outside of the plasma is used for the generation of the reactive gas ions such as oxygen ions. Figure 6.2 shows a typical Kaufman type ion beam etching system.

The ion beam sputter process achieves an anisotropic etching pattern while isotropic etching is obtained by conventional chemical wet etching. Typical etching rates in the ion beam sputter are listed in Table 6.1 (4). Photoresists are used for a mask pattern. A directional etching including vertical etching in a trench structure is achieved by the ion beam sputter process. It is noted that the etching rate strongly depends on the incident angles of the ion beam as shown in Fig. 6.3. Suitable incident angles should be selected when the photoresists are used for the mask pattern.

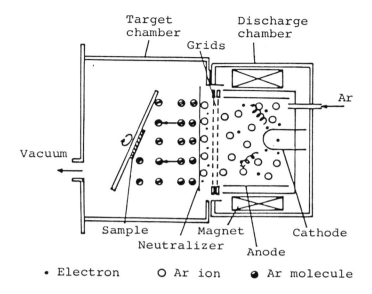

Figure 6.2: Schematic diagram of an ion-beam etching apparatus with a neutralizer. Ions generated in a discharge chamber are collimated by grids and collide to etch the sample in the target chamber.

Target Material	Composition	Etch Rate (Å/ min)
Silver	Ag	1400
Aluminum	Al	520
Alumina	Al_2O_3	100
Gold	Au	1400
Beryllium	Be	180
Bismuth	Bi	8500
Carbon	C	50
Cadmium Sulphide	CdS (1010)	2100
Cobalt	Co	510
Chromium	Cr	530
Copper	Cu	820
Dysprosium	Dy	1050
Erbium	Er	880
Iron	Fe	380
Iron Oxide	FeO	470
Gallium Arsenide	GaAs (110)	1500
Gallium Gadolinium Garnet	GaGd	280

Table 6.1: Etching rates by ion beam sputtering with Ar^+ at 500 eV and $1mA/cm^2$ (Bunshah (1982) (1)).

Target Material	Composition	Etch Rate (Å/ min)
Gallium Phosphide	GaP (111)	1400
Gallium Antimonide	GaSb (111)	1700
Gadolinium	Gd	1000
Hafnium	Hf	590
Indium Antimonide	InSb	1300
Iridium	Ir	540
Lithium Niobate	$LiNbO_3$ (Y-cat.)	400
Manganese	Mn	870
Molybdenum	Mo	230
Niobium	Nb	390
Nickle	Ni	500
Nickle Iron	NiFe	500
Osmium	Os	450
Lead	Pb	3100
Lead Telluride	PbTe (111)	3000
Palladium	Pd	1100
Platinum	Pt	780
Rubidium	Rb	4000
Rhenium	Re	470
Rhodium	Rh	650
Ruthenium	Ru	580
Antimony	Sb	3200
Silicon	Si	370
Silicon Carbide	SiC (0001)	320
Silicon Dioxide	SiO_2	400
Samarium	Sm	960
Tin	Sn	1200
Tantalum	Ta	380
Thorium	Th	740
Titanium	Ti	320
Uranium	U	660
Vanadium	V	340
Tungsten	W	340
Yttrium	Y	840
Zirconium	Zr	570
Resists	AZ 1350J	300
	COP	800
	PBS	900
	KTFR	290
	PMMA	560
	Riston 14	250
	Kodak 809	320
Glass (Na,Ca)		200
Stainless Steel		250

Table 6.1: (cont.) Etching rates by ion beam sputtering at 500 eV, Ar+, 1mA/cm² (Bunshah (1982) (1)).

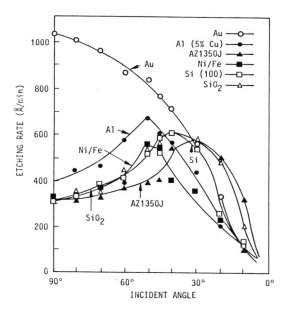

Figure 6.3: Etching rate vs. incident angle of argon ion beam (Veeco (4)).

It is widely known that the mask pattern is often eroded and then an ideal vertical etching is not achieved. The sputter etched structure shows the tapered edge. In some cases a trenching structure appears in a bottom of the etched pattern as shown in Fig. 6.4 (5). The formation of the trenching structure will result from the deposited materials sputtered from the side wall of the etched groove.

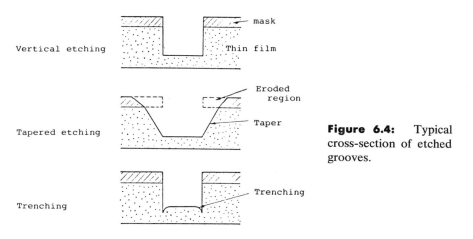

Figure 6.4: Typical cross-section of etched grooves.

This type of sputter etching is useful for a microfabrication of thin films of alloys or compounds. Thin films of compounds PLZT described in Chap. 5, for instance, are etched by the ion beam sputter system and thin film optical channel waveguides are successfully fabricated. A typical construction of the optical channel waveguides is shown in Fig. 6.5 (6).

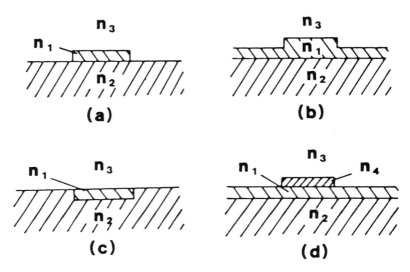

Figure 6.5: Cross-sectional view of four kinds of channel waveguide; (a) raised-strip type, (b) ridge type, (c) embedded type, and (d) strip-loaded type; n_1, n_2, n_3, n_4 are refractive indices of waveguide, substrate, environments, and loaded strip, respectively.

The channel waveguides are classified into four types: (a) raised- strip type, (b) ridge type, (c) embedded type, and (d) strip loaded type. The optical index at the inside of the optical channels is designed to have higher values compared with the optical index at the outside of the optical channels.

The thin films of the PLZT are epitaxially grown on the sapphire substrate by the sputtering deposition. These epitaxial PLZT thin films exhibit both excellent optical transparency and strong electrooptic effects.

The channel waveguide pattern of the ridge type is drawn on the PLZT thin films with a 0.4μm thick photoresist. The ion beam etching conditions are shown in Table 6.2. Etching rates of the PLZT thin films are 130Å/ min in argon atmosphere. The ratio of the etching rate PLZT/photoresist (AZ1400) is around 1.2 at the ion beam acceleration voltage of 550V, whereas the photoresist pattern is deformed at the ion beam acceleration voltage beyond 750 V. The waveguide patterns are etched at incident ion beam angles of 30° to obtain tapered ridge walls. The atomic ratio of lead to titanium is not influenced by the waveguide fabrication process.

Ar ion current	600 μA/cm^2
Acceleration voltage	550 V
Incident angle	30°
Argon pressure	1×10^{-4} Torr
Etching rate of PLZT thin films	130 A/min
Etching rate ratio PLZT thin films/AZ1400	1.2

Table 6.2: Ion beam etching conditions for PLZT thin films.

Figure 6.6 shows a typical photograph of the PLZT thin film channel waveguide. A curved optical channel waveguide is also successfully provided by the ion beam sputtering process. Typical curved waveguide is shown in Fig. 6.7. These channel waveguides will be useful for making thin film integrated optical devices.

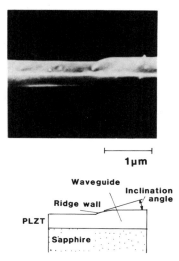

Figure 6.6: Cross-sectional views of thin film optical channel waveguide.

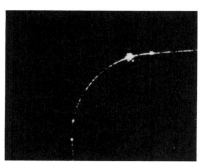

Figure 6.7: Photographs of curved optical channel waveguide: line width 10μm, curvature r = 1mm.

The ion beam sputter etching is also useful for making microstructures of the high T_c superconducting thin films (7,8). The sample is held on a sample table and tilted with respect to the direction of the incident ion beam. Typical etching conditions are shown in Table 6.3.

Applied voltage/current	550 V/600 μA/cm²
Argon pressure	1.3×10^{-2} Pa
Incident angle	45 - 90°
Sample table	Water cooling (10°C)
Sample	High-Tc superconducting films
	(Er-Ba-Cu-O, Ga-Ba-Cu-O)
Photoresists	OMR87 (negative)
	MP1400 (positive)
Etching rates	Superconductors: 250 ∿ 300 A/min
	Resists: 200 ∿ 300 A/min

Table 6.3: Ion beam etching conditions for high T_c superconducting thin films.

The high T_c Gd-Ba-Cu-O thin films are prepared by rf-magnetron sputtering on (100) MgO single-crystal substrates. The film thickness is around 5,000Å. Figure 6.8 shows typical etching rates by the argon ion bombardments for the high T_c superconducting films and the photoresists. The etching rates of the high T_c superconducting films are higher than those of the photoresists at the incident angle over 45 degrees. This suggests that the high T_c superconducting films can be patterned by the argon ion beam etching with a photoresist etching mask of about the same thickness as the high T_c thin film.

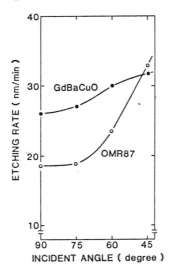

Figure 6.8: Etching rates of Gd-Ba-Cu-O superconducting film and negative-type photoresist, OMR87, as a function of the incident angle.

The procedure for patterning the film is as follows: The OMR87 resist is spin coated onto the high T_c thin film and prebaked at 95°C for 30 minutes. The resist of about 1 μm thickness is irradiated with UV light by the contact printing method and then developed. After the postbake of the resist at 95°C for 1 hour, the high T_c thin film is etched by the argon ion beam etching for 25 minutes at 60° incident angle with the etching mask of the patterned resist. The resist is finally removed by dissolving it in organic liquid, 1,1,1-trichloroethane.

Figure 6.9 shows typical resistivity-temperature curves for the microstrip lines patterned by the ion beam etching process. The resistance of the 10 μm strip approximately shows the same temperature dependence as that of the initial film before the ion beam etching, although the zero-resistance temperature slightly decreases for the 2 μm strip high T_c thin films. This suggests that the patterning of the high T_c superconducting thin films is achieved by the ion beam etching without any post-heat-treatments. Generally the post-heat-treatments are necessary after a microfabrication of the high T_c superconducting thin films when the wet etching is conducted for a pattern formation.

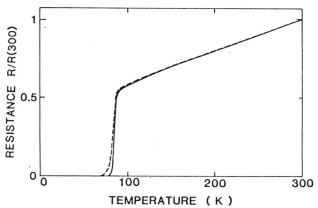

Figure 6.9: Temperature dependence of the resistance of the Gd-Ba-Cu-O film before being patterned, solid line, and that of the 2μm wide strip line, broken line. The resistance is normalized to that at 300 K.

The ion source of Kaufman type is not useful for production of high ion current. The plasmatron type ion source comprising the arc discharge can provide a higher current than the Kaufman type ion source. An additional magnetic field is imposed in order to have higher ion current.

ECR (electron cyclotron resonance) type cold cathode discharge is used for the high current ion source. The ECR discharge is sustained under an rf-electric field with static magnetic field. The ECR conditions are given by where f denotes the frequency of the rf-electric field, B, magnetic field strength, e and m electron charge and mass. For the f=2.45 GHz, the B becomes 874 G.

A typical ion beam etching system is shown in Fig. 6.10. The system is not compact since the system requires large water-cooled magnets. However, the system is suitable for the use of chemically reactive gases, since the ECR-type ion source has no hot filaments. The operating pressure is as low as 10^{-5} Torr so the ions sputter the surface of the test pieces without any gas phase collisions. This achieves the vertical etching and/or directional etching as shown in Fig. 6.11 (Anelva (9)).

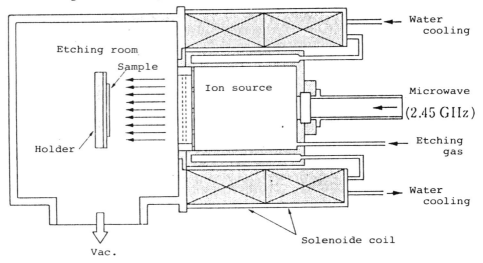

Figure 6.10: Construction of an ECR ion beam etching system.

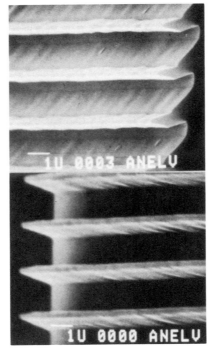

Figure 6.11 Etching structures SiO_2 on Si prepared by ECR ion beam etching system (9).

6.2 DIODE SPUTTER ETCHING

When the test samples are placed on a surface of the cathode in the diode sputtering system, the surface of the samples is etched by the incident ions. This kind of the sputter etching system is very compact, although there are some demerits such as a contamination of the samples due to the residual gas and the temperature rise of the samples during the sputter etching.

It is noted that the diode sputter etching system is useful for the reactive ion etching, RIE, since the chemically reactive gas such as oxygen and/or halogen gas could be used as the sputter gas.

Table 6.4 shows the typical etching rates for various materials in the diode RIE system (10). These etching systems are widely used for a microfabrication in the LSI, since the narrow patterns are obtained as shown in Fig. 6.12.

Materials	Etching rate (Å/min)	Etching ratio Materials/ posi- photoresists	Etching ratio Materials/ SiO_2	Etching gas
Al	1,000	6	13	$BCl_3 + CCl_4$
Poly-Si	1,700	4	18	$CCl_4 + He + O_2$
Cr	200	1		$CCl_4 + O_2$
Mo	4,000	4	100	$CCl_4 + O_2$
GaAs	6,000	6		$CCl_4 + O_2$
SiO_2	600	3	9*	CHF_3
SiN_3	600	3	9*	CHF_3
PSG**	1,200	6	18*	CHF_3
TaN	120	0.5		$CF_4 + O_2$
TaSi	1,000	1		$CF_4 + O_2$
Ti	500	1	10	CF_4
WSi	2,000	2		$SF_6 + O_2$
InSb	300	1		Ar
Polyimide	2,000		100	O_2

*   On Si   (etching power $0.25 \ \text{W/cm}^2$)
**   8 mol % P doped PSG

Table 6.4: Etching rates of RIE for various materials (Anelva (10)).

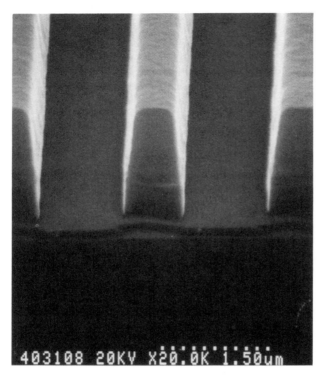

Figure 6.12: Narrow etched structure SiO₂ on poly-Si prepared by RIE at diode sputter system (Anelva (9)).

It is also noted that the RIE process achieves a micro-etching of the chemically stable materials such as diamonds. Diamond is a high temperature stable material. It is stable as high as 1700°C in vacuum, and 600°C in air. The surface is conventionally etched by a molten potassium nitrate at around 700 − 800°C. Reactive ion etching with oxygen results in surface etching even at a room temperature. Figure 6.13 shows a typical etched surface of the diamond (110) crystals (11-13).

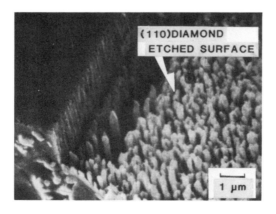

Figure 6.13: Etched surface of (110) diamond crystal treated by oxygen sputtering.

6.3 PLASMA ETCHING

The diode sputter etching system is used for the plasma etching process. For the plasma etching process a high density reactive gas is introduced. The reactive gas molecules excited by a hot electron in the plasma etch the surface of the specimens. The plasma etching is isotropic and shows less irradiation damage than the sputter etching process.

There are several plasma etching systems. Typical systems are shown in Fig. 6.14 (14). Wide varieties of materials such as Al, Au, Cr metals, Si, GaAs semiconductors, and SiO_2, Si_3N_4 dielectrics are successfully etched by the plasma etching process. Table 6.5 shows the typical plasma etching conditions for various materials.

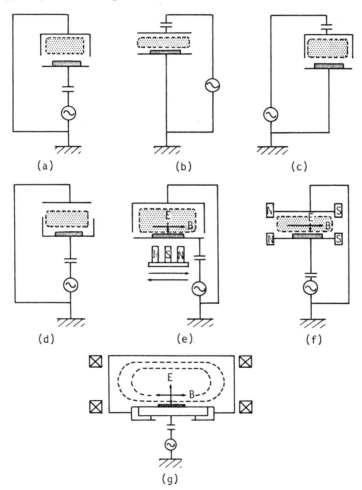

Figure 6.14: Typical constructions of various plasma etching systems (Anelva (14)).

MATERIALS	Al	Au	Cr	CrOx	CrOx (RIE)	Fe-Ni	InOx	Mo	Ni-Cr
SUBSTRATE	SiO$_2$ Si$_3$N$_4$ PSG	Ceramic	Glass	Cr Glass	Glass	Ceramic	Glass	SiO$_2$	Ceramic
BASE PRESSURE	$\leq 7\times10^{-3}$ Pa $(5\times10^{-5}$ Torr)	$\leq 4\times10^{-2}$ Pa $(3\times10^{-4}$ Torr)	$\leq 4\times10^{-2}$ Pa $(3\times10^{-4}$ Torr)	$\leq 4\times10^{-2}$ Pa $(3\times10^{-4}$ Torr)	$\leq 4\times10^{-2}$ Pa $(3\times10^{-4}$ Torr)	$\leq 4\times10^{-2}$ Pa $(3\times10^{-4}$ Torr)	$\leq 4\times10^{-2}$ Pa $(3\times10^{-4}$ Torr)	$\leq 4\times10^{-2}$ Pa $(3\times10^{-4}$ Torr)	$\leq 4\times10^{-2}$ Pa $(3\times10^{-4}$ Torr)
ETCHING GAS I (DEM-451)	CCl$_4$ 150 c.c./min (16 Pa, 0.12 Torr)	CCl$_2$F$_2$ 50 c.c./min (13 Pa, 0.10 Torr)	CCl$_4$ 150 c.c./min (16 Pa, 0.12 Torr)	CCl$_4$ 85 c.c./min (13 Pa, 0.10 Torr)	CCl$_4$ 150 c.c./min (16 Pa, 0.12 Torr)	CCl$_4$ 50 c.c./min (10 Pa, 0.075 Torr)	CCl$_4$ 85 c.c./min (13 Pa, 0.10 Torr)	CF$_4$ 100 c.c./min (13 Pa, 0.10 Torr) / CCl$_4$ 150 c.c./min (16 Pa, 0.12 Torr)	CCl$_4$ 50 c.c./min (10 Pa, 0.075 Torr)
ETCHING GAS II (DEM-451)	—	—	Air 150 c.c./min (24 Pa, 0.18 Torr)	—	Air 150 c.c./min (24 Pa, 0.18 Torr)	O$_2$ 10 c.c./min (2 Pa, 0.015 Torr)	—	O$_2$ 10 c.c./min (1.3 Pa, 0.01 Torr) / O$_2$ 30 c.c./min (3.2 Pa, 0.024 Torr)	O$_2$ 10 c.c./min (2 Pa, 0.015 Torr)
GAS PRESSURE	8 Pa (0.06 Torr)	4 Pa (0.03 Torr)	40 Pa (0.3 Torr)	4 Pa (0.03 Torr)	40 Pa (0.3 Torr)	4 Pa (0.03 Torr)	4 Pa (0.03 Torr)	40 Pa (0.3 Torr)	4 Pa (0.03 Torr)
RF-POWER DEM-451	150 W	200 W	150 W	300 W	150 W	300 W	300 W	50 W	300 W
RF-POWER DEA-503	300 W	400 W	300 W	600 W	300 W	600 W	600 W	100 W	600 W
POWER DENSITY	0.25 W/cm²	0.33 W/cm²	0.25 W/cm²	0.5 W/cm²	0.25 W/cm²	0.5 W/cm²	0.5 W/cm²	0.08 W/cm²	0.5 W/cm²
ELECTRODE SPACING DEM-451	40~80 mm	40~80 mm	20~60 mm	40~80 mm	20~60 mm	40~80 mm	40~80 mm	20~60 mm	40~80 mm
ELECTRODE SPACING DEA-503	60~100 mm	60~100 mm	40~80 mm	60~100 mm	40~80 mm	60~100 mm	60~100 mm	40~80 mm	60~100 mm
END MONITOR	voltage opt.		voltage		voltage			voltage	
ETCHING RATE	2000 Å/min	200 Å/min	150 Å/min	100 Å/min	100 Å/min	100 Å/min	150 Å/min	1500 Å/min	200 Å/min
SELECT BASE	30	10	10	Cr Glass 5 4	5	5	5	100	10
SELECT RESIST(posi)	4	—	15	—	1	—	—	10	—

Table 6.5: Typical plasma etching conditions for various materials.

In the plasma etching process, the reacted products from the surface of the sample are vaporized during the etching. However, the products may remain on the etched surface when the reacted products is not volatile at the temperature of the sample. A critical example of this effect is etching of Cu by $CFCl_3$ (15).

It is noted that a low temperature inter diffusion is often observed in a multi-layer system during the plasma etching process. Figure 6.15 shows a typical thickness distribution of compositions in a layered thin film structure of Cu thin film/Cr thin film/glass substrate after the plasma etching by CCl_2F_2 Cr and/or Si out-diffuse onto the surface of the Cu even at the temperature of 120°C The Si may out-diffuse from the glass substrates.

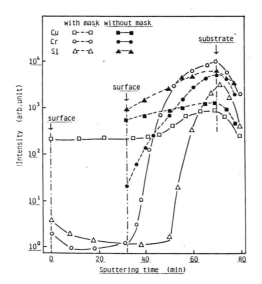

Figure 6.15:

The ion beam sputter etching, diode sputter etching, and the RIE are dispensable for the microfabrication, although the etching mechanism is not wholly understood yet.

6.4 REFERENCES

1. Bunshah, R.F. (ed.), Deposition Technologies for Films and Coatings, New Jersey, Noyes Publications (1982).

2. Vossen, J.L., and Kern, W., (ed.), Thin Film Processes, New York, Academic Press (1978).

3. Wilson, R.G. and Brewer, G.R., Ion Beams, New York, John Wiley & Sons (1973).

4. Veeco, catalogue, No. E-0029.

5. Chapman, B., Glow Discharge Processes, New York, John Wiley & Sons (1980).

6. Kawaguchi, T., Adachi, H., Setsune, K., Yamazaki, O., and Wasa, K., Applied Optics, 23: 2187 (1984).

7.   Enokihara, A., Higashino, H., Setsune, K., Mitsuyu, T., and Wasa, K., <u>Jpn. J. Appl. Phys.</u> 27: L1521 (1988).

8.   Higashino, K., Enokihara, A., Mizuno, K., Mitsuyu, T., Setsune, K., and Wasa, K., <u>Proc. FED HiTcSc-ED Workshop,</u> June, 1988, Miyagi-ZaO, P.267.

9.   Anelva catalogue, No. 154-14J.

10.   Anelva catalogue, No. 154-01J.

11.   Wasa, K., <u>National Tech. Rept.</u> 22: 836 (1976).

12.   Hayakawa, S., and Wasa, K., <u>Thin Film Technology,</u> p. 67, Tokyo, Kyoritsu (1982), Wasa, K., and Hayakawa, S., <u>Sputtering Technology,</u> p. 209, Tokyo, Kyoritsu (1988).

13.   Efremow, N.N., Geis, M.W., Flanders, D.C., Lincoln, G.A., and Economow, N.P., <u>J. Vac. Sci. Technol.,</u> B3: 416 (1985).

14.   Anelva News, No. 25 (1986).

15.   Kawaguchi, T., and Wasa, K., <u>Oyo Butsuri</u> 51: 856 (1982).

7

FUTURE DIRECTIONS

7.1 CONCLUSIONS

Sputtering was first observed over 140 years ago in a discharge tube by Bunsen and Grove. Since that time, the basic level of understanding of the sputtering process has become fairly well developed. The applications of sputtering, however, are still being developed on a daily basis. Sputtering and sputter deposition have become common manufacturing processes for a wide variety of industries. First and foremost is the semiconductor industry, in which virtually every integrated circuit produced uses sputtering technology in some form. This book has examined many of the sputtering applications that are relevant to IC fabrication and applications.

Sputtering and sputter deposition are also present in many other disparate areas. For example, sputter deposition is used to coat the mirror-like windows in many tall buildings. It is also used in the food processing industry to coat the plastic packaging used for such things as peanuts and potato chips. A wide range of applications have been developed for the automobile industry, such as coating reflective surfaces onto plastics. A totally different application is the sputter deposition of hard, resilient coatings onto drill bits and cutting tools. These very hard, wear-resistant coatings can increase the lifetimes of drills and punches by orders of magnitude. In addition, many of these same sputter-deposited films have application as decorative layers, on such items as watch casings and eyeglass frames. Unfortunately, it has been beyond the scope of this book to cover each of these applications, many of which are entire industries unto themselves.

Sputtering, by itself, is an invaluable tool for the analysis of surfaces. By using a surface sensitive analysis technique, such as Auger Electron Spectroscopy (AES), the surface can be slowly removed by sputtering and a depth profile of the near surface can be obtained. A much different application of sputtering occurs in the actual machining of parts by ion bombardment. An example of this is present in the magnetic heads used on most disk drives.

Sputter deposition is unique, compared to other techniques of film preparation, in that sputter deposition is a quenched, high energy process. Films deposited by other techniques, such as evaporation or chemical vapor deposition, are formed under conditions of thermodynamic equilibrium. This tends to produce materials with qualities much like bulk materials which are synthesized by routine melting or sintering. During sputter deposition, the depositing atoms are quite energetic, compared to the substrate temperature. Upon deposition, the atom's kinetic energy is deposited locally on the substrate surface, which then cools rapidly. This dynamic, quenching process allows the formation of novel materials which may not have been formed under conventional, thermal equilibrium conditions. The novel properties may include new stoichiometries, unusual crytalline orientations (or lack of orientation), enhanced bonding of the film material to the substrate, and even metastable properties which are quenched into the bulk of the film. By adjusting the relative energies and fluxes of the atom and ion species present during sputter deposition, even more control can be exerted on the dynamics of the film formation process. The area of bombardment-enhanced film modification has only recently become recognized as a major new topic in the realm of materials science.

Sputter deposition has also led to an almost atom-by-atom construction of new, important materials. For example, the high temperature superconductors described in this book are critically sensitive to the relative ordering of the arriving atoms, as well as the temperature and orientation of the substrate. By sputtering, it is possible to form atomic layers of the correct materials, which can then be processed into the correct phase and oxidation level to produce the high temperature superconducting effect.

Many areas are still open for exploration in the realm of sputtering and sputter deposition. For example, low energy sputtering (below 100 eV) plays a critical part in the new generation of very dense plasma machines (1). For example, in an ECR plasma source, the net ion current to the chamber walls may be in the 10 ampere range. With even an infinitesimal sputter yield (e.g., 10^{-6}, 4 orders of magnitude below the currently recognized thresholds), the erosion rate of the walls can be on the order of a monolayer per minute, resulting in significant metallic contamination of sensitive IC parts (1).

The effect of low energy ion bombardment during deposition is only recently being quantified. As mentioned above, these systems are not in thermodynamic equilibrium. This suggests that changing the incoming ion energy without changing the total incoming power (by adjusting the flux) will result in different film properties. This is critically important to such topics as grain growth and orientation, the inclusion of gas within the lattice, and film stress and adhesion.

Sputter deposition has moved beyond the realm of a purely empirical, experimental discipline. New computer models are being developed which model the sputtering process following each atom in the lattice, and representing the lattice in a realistic mode similar to a solid (2). Models are also being refined which examine the dynamics of film growth by sputtering (3). These models are invaluable in discerning the effect of the sputtered atom's energy and direction, as well as the effects of additional incident particles and sample temperature.

One of the most promising areas of sputter deposition technology is the area of multilayer depositions. These artificial superlattices can have novel properties by themselves, or be used in the formation of new chemical structures. For example, silicides can be formed under very controlled conditions by the alternate deposition of many layers of silicon and an appropriate metal and subsequent heat treatment.

Other new areas of sputtering research have focussed on the modification of the directionality of the sputtered atoms. For example, the imposition of a physical collimator near a sputtering target results in a very directional flux of atoms which can be used for a variety of processing steps with ICs (4). Low pressure sputter deposition is a valuable area, because it removes the confusing effects of gas collisions on the film properties.

Low pressure sputtering also allows the use of multiple sources or plasmas within a single chamber. The long mean free paths of the sputtered atoms are useful in depositing novel alloy materials.

Sputtering will continue to be a growing field, in terms of applications, as well as in the area of basic studies. New materials and applications of these materials are being developed daily and many of the problems associated with sputtering in the past are being eliminated. Sputtering technology is becoming commonplace in many manufacturing disciplines, and new areas are emerging.

7.2 REFERENCES

1. Rossnagel, S.M., in preparation (1990).

2. Ruzic, D.N., in: The Handbook of Plasma Processing Technology, (S.M. Rossnagel, J.J. Cuomo and W.D. Westwood, eds.) chap. 3, p.70, Noyes Publications, Park Ridge, NJ (1990).

3. Muller, K-H., in: Handbook of Ion Beam Processing Technology, (J.J. Cuomo, S.M. Rossnagel and H.R. Kaufman, eds.), p. 241, Noyes Publications, Park Ridge, NJ (1989).

4. Rossnagel, S.M., Kinoshita, H., Mikelson, D., and Cuomo, J.J., J. Vac. Sci. & Technol., A9: (in press).

APPENDIX

Electric Units, Their Symbols and Conversion Factors*

Physical quantity		Name of electro-magnetic unit	Name of practical unit	Name of M.K.S. unit	M.K.S. unit Electro-magnetic unit	M.K.S. unit Practical unit
Force		dyn	joule/cm	newton (joule/m)	10^5	$1/10^2$
Energy		erg	joule	joule	10^7	1
Power		erg/s	watt	watt	10^7	1
Potential	V		volt	volt		1
Electromotive force	E		volt/cm	volt/m		$1/10^2$
Current	I		ampere	ampere		1
Resistance	R		ohm	ohm		1
Resistivity	ρ		ohm-cm	ohm-m		10^2
Conductance	G		mho	mho (siemens)		1
Conductivity	κ		mho/cm	mho/m		$1/10^2$
Electric charge	q, Q		coulomb	coulomb		1
Electric displacement	D			coulomb/m^2		$4\pi/10^4$
Electric polarization	P			coulomb/m^2		$1/10^4$
Capacitance	C		farad	farad		1
Permitivity of free space	ε_0			farad/m		$4\pi/10^9$
Magnetomotive force		gilbert	ampere-turn	ampere-turn	$4\pi/10$	1
Magnetic field strength	H	oersted	$\dfrac{\text{ampere-turn}}{\text{cm}}$	$\dfrac{\text{ampere-turn}}{\text{m}}$	$4\pi/10^3$	$1/10^2$
Magnetic flux	ϕ	maxwell	maxwell	weber	10^8	10^8
Magnetic flux density	B	gauss	gauss	weber/m^2 (tesla)	10^4	10^4
Intensity of magnetization				weber/m^2	$10^4/4\pi$	$10^4/4\pi$
Magnetic charge	M			weber	$10^3/4\pi$	$10^8/4\pi$
Magnetic moment	I			weber-m	$10^{10}/4\pi$	$10^{10}/4\pi$
Inductance	L		henry	henry	10^9	1
Magneto resistance				$\dfrac{\text{ampere-turn}}{\text{weber}}$	$4\pi/10^9$	$4\pi/10^9$
Permeability of free space	μ_0			henry/m	$10^7/4\pi$	$10^7/4\pi$
Magnetic susceptibility	X			henry/m	$10^7/(4\pi)^2$	$10^7/(4\pi)^2$

* example (force): $1 \text{ dyn} = 10^{-5} \text{ N} = 10^2 \text{ joule/cm}$

Fundamental Physical Constants

Velocity of light	$c = 2.998 \times 10^8$ m·sec^{-1}, $- \times 10^{10}$ cm·sec^{-1}
Electron rest mass	$m_c = 9.11 \times 10^{-31}$ kg, $- \times 10^{-28}$ g
Proton rest mass	$m_p = 1.67 \times 10^{-27}$ kg, $- \times 10^{-24}$ g
Charge of electron	$e = 1.602 \times 10^{-19}$ C, 4.8×10^{-10} esu, 1.602×10^{-20} emu
Charge-to-mass ratio of electron	$e/m_c = 1.76 \times 10^{11}$ C·kg^{-1}, 5.27×10^{17} esu·g^{-1}, 1.76×10^7 emu·g^{-1}
Planck's constant	$h = 6.626 \times 10^{-34}$ J·sec, $- \times 10^{-27}$ erg·sec
	$h/2\pi = \hbar = 1.055 \times 10^{-34}$ J·sec, $- \times 10^{-27}$ erg·sec
Bohr radius	$= 5.29 \times 10^{-11}$ m $- \times 10^{-9}$ cm
Bohr magneton	$= 9.27 \times 10^{-24}$ J·tesla^{-1}, $- \times 10^{-21}$ erg·gauss^{-1}
Magnetic moment	
Electron	$= 9.28 \times 10^{-24}$ J·tesla^{-1}, $- \times 10^{-21}$ erg·gauss^{-1}
Proton	$= 1.41 \times 10^{-26}$ J·tesla^{-1}, $- \times 10^{-23}$ erg·gauss^{-1}
Boltzmann's constant	$k = 1.38 \times 10^{-23}$ J·K^{-1}, $- \times 10^{-16}$ erg·K^{-1}
	$= 8.615 \times 10^{-5}$ eV·K^{-1}
Avogadro's number	$= 6.02 \times 10^{23}$ mole^{-1}
Volume of 1 mole of an ideal gas	$= 2.24 \times 10^{-2}$ m^3·mole^{-1}, $- \times 10^4$ cm^3·mole^{-1}
Loschmidt's constant	$= 2.69 \times 10^{25}$ m^{-3}, $- \times 10^{19}$ cm^{-3}
Gas constant	$R = 8.31$ J·mole^{-1}·K^{-1}, $- \times 10^7$ erg·mole^{-1}·K^{-1}, 1.99 cal·mole^{-1}·K^{-1}
Faraday's constant	$= 9.65 \times 10^4$ C·mole^{-1}, 2.89×10^{14} esu·mole^{-1}, 9.65×10^3 emu·mole^{-1}
Permittivity of free space	$\epsilon_O = 8.854 \times 10^{-12}$ C·V^{-1}·m^{-1}
Permeability of free space	$\mu_O = 1.26 \times 10^{-6}$ H·m^{-1}
Standard value of the acceleration of gravity	$g = 6.67 \times 10^{-11}$ N·m^2·kg^{-2}, $- \times 10^{-8}$ dyne·cm^2·g^{-2}
Length	$1\,\mu = 10^{-6}$ m, 10^{-4} cm
	$1\,\text{Å} = 10^{-10}$ m, 10^{-3} cm
	1 mil $= 2.54 \times 10^{-5}$ m, $- \times 10^{-3}$ cm
Work	1 dyne $= 10^{-5}$ N
	1 g·wt $= 9.81 \times 10^{-3}$ N, $- \times 10^2$ dyne
Atmospheric pressure	1 Torr $= 1$ mm Hg, 1.33 mbar, 1.33×10^2 Pa
	1 atm $= 760$ Torr, 1013 mbar
Temperature	$0°$C $= 273.15$ K
Energy	1 erg $= 10^{-7}$ J
	1 cal $= 4.18$ J
1 eV	Wavelength $= 1.24 \times 10^{-6}$ m, $- \times 10^{-4}$ cm
	Frequency $= 2.42 \times 10^{14}$ sec^{-1}
	Wave number $= 8.07 \times 10^5$ m^{-1}, $- \times 10^3$ cm^{-1}
	Energy $= 1.60 \times 10^{-19}$ J, $- \times 10^{-12}$ erg
	Temperature $= 1.16 \times 10^4$ K

INDEX

abnormal glow	83
acceptor	260
acoustic properties	150
activated reactive evaporation	25
activation energy	164
adatoms	32
adhesion	18, 42
adsorption	10
Ag	38
Al	124
AlN	39,124,264
Al_2O_3	124,126,234,264
amorphous	157,172,199,248
amorphous Si	6
anisotropy	198
annealing	14
asymmetric discharge	96
Au	12,38,124
$Au_{.19}A$	64
Au_2Al	64
Auger electron spectroscopy	37
auxiliary cathodes	112
B_4C	234
$BaTiO_3$	124
$Bi_{12}GeO_{20}$	126,156
binding energy	75
Bi_2O_3	160
$Bi_{12}PbO_{19}$	127
birefringence	190
$Bi_{12}SiO_{20}$	156
Bi-Sr-Ca-Cu-O	132,193,213
$Bi_{12}TiO_{20}$	40,126
Born-Mayer potential	76
Born's approximation	73
Bragg angle	155

296

Bragg diffraction   155
Bravis Empirical Law   135, 258
BTO   40
buffer layer   146

carbides   231
cathode dark space   84
cathode fall   86
CdS   16,38,124
chemical vapor
   deposition   19, 144, 198
cluster   10, 26
cluster deposition   144
cluster emission   65, 77
Co   38,124
coalescence   11
cold-cathode discharge   81
cold plasma   95
columns   16
compound sputtering   110
computer simulation   74
condensation   13
cosine law   57
co-sputtering   259
coupling constant   189
coupling factor   147
Cr   124
Cr-Si   241
critical current   198,205,217,222
critical magnetic field   5
critical nuclei   10
Crookes dark space   84
crystal structure   59
crystallization temp.   32,167,204,
                               216,249,253
Cu   38,124
Cu$_3$Al   64
CuO   210
Curie Temperature   186
cycloidal motion   89
cylindrical magnetron   97

dark-current-voltage   247
dc diode   23, 97
deep trap   260
defects   12
density   17

deposition rate   98
devices   6
diamagnetic   218
diamagnetization   221
diamond   2,231, 242, 286
dielectric   252
dielectric constant   44, 171
diffusion   210
diode   285
dislocation   13
dissociation   28
DLC   242
donor   260
doping   259
duoplasmatron   105

ECR   69,106,144,
                              198,223,283
ejection direction   60
electrical resistivity   144
electromechanical
   coupling   149, 173, 189
electron density   117
electron mean free path   94
electron mobility   95
electron screening   72
electron temperature   27,117,121
electro-optics   2,190
ellipsometry   41,190
epitaxial temperature   146,169,176
epitaxy   12,15,33,144,160,
                              168,199,208,246
Er-Ba-Cu-O   210
Er$_2$O$_3$ 210
evaporation   20,29
excitation   28
ex-situ annealing   204

Fe   38,124
ferroelectric materials   250
field effects   4
forbidden gap   260
Frank-van-der
   Merwe type   11
friction   44

GaAs   39
GaN   39
GaP   39,64
Gd-Ba-Cu-O   203,282
Ge   38
GeO$_2$ 156
GeSe   16
GeSi   64
glow discharge   83
grain size   14
growth   10

Hall measurements   225
hardness   43
hemispherical sputtering   142
high frequency discharge   87
high temp.
    superconductors   125,193
hollow cathode   276
hot plasma   95
hysteresis   222

impedance matching
    network   99
impurities   13
in-situ annealing   204
in-situ deposition   204,216
incident ion angle   57
inclusion   25
infrared absorption   230
In$_2$O$_3$ 124,226
InP   64
InSb   39
integrated optics   6
interatomic distance   60
interdiffusion   183,289
interferometer   41
ion beam sputtering   25,105,230,275
ion energy   50,97
ion plating   25, 144
ionization   28,83
ionization coefficient   84
ionization efficiency   94
ionization energy   94
ionized cluster beam   26
island structure   2

Johnson noise   175

Kaufman ion source   105,276
Kerr effects   190
K_2NiF_4 structure   193

La-Sr-Cu-O   203
Langmuir probes   117
laser assisted CVD   29
laser deposition   21
lattice defect   3
lattice mismatch   160,177
layer-by-layer deposition   220
Lenz Law   222
Lichteneker's
  empirical law   163
$LiNbO_3$ 40,251
$LiTaO_3$ 251
liquid sputtering   78

magnetic anisotropy   4
magnetic field   87
magnetic heads   6
magnetron   24,97,100
                            177,257
masking   275
Mason equivalent circuit   173
Maxwellian   120
metastable   17
microhardness   233
microstructure   13
microtwins   13
mobility   3
molecular beam epitaxy   21
monolayer   50
MoSi   241
multiple-positioning
  boundaries   13
multiple target
  sputtering   181

NaCl   12
Nb   254
NbN   254
Nd-Ce-Cu-O   225
negative bias   95,99
negative space charge   93
NiCr   62
NiSi   64
nitrides   227

normal glow   83
nucleation   10
nucleation barrier   16

onset temperature   205,224
optical absorption   260
optical band gap   18
optical interferometer   122
optical switches   191
organic thin films   256
orientation   12,15,139
oscillator   151
over-cosine   58
oxidation   140
oxygen vacancy   203

partial ionization   65
Pashen's law   84
passivation   223
patterning   283
Pb   176,253
PbO   156,253,262
PbTiO$_3$ 40,162,251,262
Pb$_2$Ti$_2$O$_6$ 166
permalloy (81Ni-19Fe)   62
permitivity   163,184
perovskites   125,162
physical vapor deposition   19
piezoelectricity   45,144,173
plasma assisted CVD   27
plasma etching   287
plasmatron   275
PLZT   40,125,175,
                                 186,280
point defects   13
polarization   188
polycrystalline   157
polymer   256
porous   17
positive column   86
positive space charge   93
postannealing   210
powders   1
power density   118
preferential ejection   57
preferred orientation   207,258
processing plasma   27

proximity effects   5
Pt$_2$Si   64
pulsed laser deposition   198
pyrochlore   166,179,198

quadrupole mass
  analyzer   117
quantum size effects   1
quartz crystal
  microbalance   50
quenching   250

radiation damage   54
Raman scattering   244
reactive gas   24
reactive ion etching   286
reactive magnetron   240
reactive sputtering   24,112,118,
                              125,162
reactor   27
reevaporation   171,182
rf diode   24
rf discharge   92
RHEED   37
roughness   15
roughness factor   16
Rutherford scattering   72

SAW   189
SAW Propagation loss   151
Sawyer-Tower circuit   186
secondary electron
  emission   54,83
selenides   245
SEM   41
shear mode coupling   147
sheath   84
Si   38
SiC   6,38,231
silicides   231
sillenites   156
SiN   223
single crystal   12,59,200
sintering   210
SiO   223
Si$_3$N$_4$ 230

SiO_2	126,156
skin effect	3
SnO_2	226
solubility	18
space charge	83
spark voltage	81
spikes	245
SPLC	246
sputtering	22,29
sputter threshold	50
sputter yield	49,52
sputter yield-alloys	61
sputtered atom energy distribution	67
sputtered atom velocity	65
sputtering cascade	70
sputtering gas	112
sputtering model	71
SQUID	221
stacking faults	13
sticking coefficient	13
STM	41
Stranski-Krastanov-type	11
stress	3,43
submicron	275
substrate temperature	116,213
superconductors	2
superconductors - high temp.	125,193
superlattices	6,21,181,254
supersaturation	15
surface acoustic wave	6,148
surface crystallinity	37
surface migration	165
surface mobility	13
surface morphology	218
Ta, tetragonal	1
TaN	227
Ta_2O_5	64
teflon	256
temp. dep. of resistivity TCR	3,210,228
tetragonal structure	211
thermal conductivity	3
thermal expansion	144
thermionic electrons	83
thermistors	236

thermoelectric power   3
thickness distribution   115
thin film heaters   228
thin film integrated optics 281
thin film resistors   228
thin film transducer   147
thin film transistors   6,238
thinning   275
Thomas-Fermi potential   76
TiN   228
TiO_2 113,126,262
Tl-Ba-Ca-Cu-O   193,214
Townsend discharge   83
transducer   147
transition temperature   205
transparent films   226
transverse magnetic field   90
trenching   279
tunneling   4,198

ultrafine particles   1
ultrasonic microscope   153
ultrasonic transducer   151

vapor pressure   157,176
vaporized   289
Volmer-Weber type   11
von Ardene ion source   105

wave guide   155,191
WC   238
wear   44
wear resistance   233,238

Y-Ba-Cu-O   198

ZnO   6,25,38,
                            134,126,258
ZnS   126
ZnSe   126,245
Zr   17,262
ZrO_2 262